ANALYSIS ON REAL AND COMPLEX MANIFOLDS

North-Holland Mathematical Library

VOLUME 35

NORTH-HOLLAND
AMSTERDAM · NEW YORK · OXFORD

Analysis on Real and Complex Manifolds

R. NARASIMHAN

University of Chicago
Chicago, IL 60637, U.S.A.

NORTH-HOLLAND
AMSTERDAM · NEW YORK · OXFORD

ISBN: 0 444 87776 2

First edition: 1968
Second printing: 1973
Third printing: 1985

(The first edition and the second printing were published as Volume 1 in the series Advanced Studies in Pure Mathematics)

Published by:
ELSEVIER SCIENCE PUBLISHERS B.V.
P.O. Box 1991
1000 BZ Amsterdam
The Netherlands

Sole distributors for the U.S.A. and Canada:
ELSEVIER SCIENCE PUBLISHING COMPANY, INC.
52, Vanderbilt Avenue
New York, NY 10017
U.S.A.

Library of Congress Cataloging in Publication Data

Narasimhan, Raghavan.
Analysis on real and complex manifolds

(North-Holland mathematical library, v. 35)
Bibliography: p.
Includes index.
1. Differentiable manifolds. 2. Complex manifolds 3. Differential operators I Title II. Series.
QA614.3 N37 1985 516 3'6 85-10155
ISBN 0-444-87776-2

Transferred to digital printing 2006

Preface

This book has its origin in lectures given at the Tata Institute of Fundamental Research, Bombay in the winter of 1964/65. The aim of the lectures was to present various topics in analysis, both on real and on complex manifolds. It is unnecessary to add that the topics actually chosen were determined entirely by personal taste. The contents were issued as lecture notes by the Tata Institute, and the present book is based on these notes.

The book is meant for people interested in analysis, who have little analytical background. The elements of the theory of functions of real variables (differential and integral calculus and measure theory) and some complex variable theory are assumed. Elementary properties of functions of several complex variables which are used are, in general, stated explicitly with references. It is however supposed that the reader is well acquainted with linear and multilinear algebra (properties of duals, tensor products, exterior products and so on of vector spaces) as well as set topology (properties of connected and locally compact spaces). (The material required is contained in Bourbaki: *Algèbre Linéaire*, *Algèbre Multilinéaire*, and *Topologie Générale*, *Chap.* I & II).

There are three chapters. The first deals with properties of differentiable functions in R^n. The aim is to present, with complete proofs, some theorems on differentiable functions which are often used in differential topology (such as the implicit function theorem, Sard's theorem and Whitneys' approximation theorem).

The second chapter is meant as an introduction to the study of real and complex manifolds. Apart from the usual definitions (differential forms and vector fields) this chapter contains an exposition of the theorem of Frobenius, the lemmata of Poincaré and Grothendieck with applications of Grothendieck's lemma to complex analysis, the imbedding theorem of Whitney and Thom's transversality theorem.

The last chapter deals with properties of linear elliptic differential operators. Characterizations of linear differential operators, due to Peetre and to Hörmander are given. The inequalities of Gårding and of Friedrichs on elliptic operators are proved and are used to prove the regularity of weak solutions of elliptic equations. The chapter ends with the approximation theorem of Malgrange–Lax and its application to the proof of the Runge theorem on open Riemann surfaces due to Behnke and Stein.

We have not dealt with Riemannian metrics and elementary differential geometry. Nor have we dealt with elliptic complexes in spite of their importance and interest. It is actually not very difficult to extend the theorems, such as the finiteness theorem of Chap. 3, to such complexes.

It remains for me to acknowledge the help I have received in preparing this book. My thanks are due to Mrs. M. Narlikar who wrote the notes issued by the Tata Institute; I am specially indebted to H. G. Diamond who read, very carefully, a large part of these notes, pointed out mistakes, and suggested improvements and different proofs. Finally, I am grateful to N. H. Kuiper for his invitation to rewrite the Tata Institute notes as a book, for his helpful remarks on Chapters 1 and 2 and for his assistance in preparing the manuscript for the printer.

Genève, July 1968.

Raghavan Narasimhan

Preface to the third printing*

The present edition of this book is simply a reprint of the second (1973) with such misprints corrected as I have noticed. I should like to take this opportunity to make a few general comments on the contents, and, for reasons given below, mention an alternative approach to the theory of linear elliptic operators (Chapter 3).

The bulk of Capters 1 and 2 consists of introductory material; without the basic facts concerning differentiable functions in R^n, or the fundamental concepts relating to manifolds, it would be difficult indeed to undertake any study of analysis on manifolds. Some of the results are, however, in a different category. As remarked in §2.15.12, Whitney's approximation theorem proved in §1.6 has an extension to arbitrary real analytic manifolds which are countable at infinity. This extended version leads to the theorem that if a real analytic manifold has a closed C^1- imbedding in R^N for some N, then it has a closed *real analytic* imbedding in the same R^N. H. Grauert, to whom this theorem is due, proved it by using his solution to the so-called Levi problem which is one of the fundamental results in one aspect of complex analysis. It should be mentioned that no purely real variable proof of this result is known.

§2.14 contains some of the main ideas in the proof of another important theorem in complex analysis, usually called "Theorem B for Stein Manifolds". The other ingredients needed in this approach are

* The references in this Preface are listed at its end.

(i) the theory of coherent analytic sheaves, and (ii) a theorem of H. Cartan on matrices of holomorphic functions.

L. Hörmander [9, 10] has developed a different method to solve the Levi problem and to prove Theorem B for Stein manifolds. He uses ideas from the general theory of linear differential operators to solve directly the equation $\bar{\partial}u = f$. This method has proved to be of great flexibility and power. Usually referred to, somewhat vaguely and incorrectly, as the $\bar{\partial}$-method, it is, for instance, one of the key ingredients in recent work relating the differential-geometric structure of a complex manifold to its analytic structure. As an important and typical paper in this work, one might cite Y. T. Siu and S. T. Yau [19].

§2.13 contains a proof of what is universally called "Poincaré's lemma": a closed differential form of degree $\geqq 1$ on R^n is exact. It was Prof. Georges de Rham who pointed out to me that Poincaré had nothing to do with this result. The facts that Prof. de Rham showed me are sufficiently interesting to record here.

Poincaré himself was never concerned with the exterior derivative at all, and never introduced it, not even under another name. What he did do was to set up, in 1895 [17], conditions under which the integral of a differential form on a "cycle depending on a parameter" is independent of the value of the parameter. Given the formalism of the exterior derivative, these conditions amount simply to saying that the form is closed. In his book [4], E. Cartan gives the name "théorème de M. Poincaré" to the essentially trivial fact that the square of the exterior derivative is zero; the local converse is called, naturally enough, "la réciproque du théorème de M. Poincaré". Cartan had already proved the local converse in an earlier book [3] without mentioning Poincaré, or anyone else for that matter.

The fact is that already in 1889, V. Volterra [21] had proved that $d^2 = 0$ *as well as the converse statement on* R^n. In fact, Volterra's papers contain a general form of Stokes' theorem (which does very efficiently somewhat more than what Poincaré did with his "cycles depending on a parameter"). They also contain the concept of "harmonic forms", a concept that has proved to be of the greatest importance. It seems a pity that Volterra gets no credit for this very original, very important work. However, the name "Poincaré lemma" has become so much a part of mathematical usage that it would probably be futile to try to change it now.

We turn now to Chapter 3. After introducing the Fourier transform and the Sobolev spaces, the chapter deals with two of the basic *à priori* estimates for linear elliptic operators, viz. the inequalities of Gårding and of Friedrichs, and with the interior regularity of weak solutions (3.6.3, 3.6.8, and 3.7.7). The rest of the chapter is concerned with transporting these results to elliptic operators between vector bundles on a manifold, and with applications. The regularity theorem is proved using a differencing argument of L. Nirenberg; this method is useful elsewhere as well (for example in the so-called $\bar{\partial}$-Neumann problem; see [7]).

I shall now describe briefly a different approach to the theory of linear elliptic operators, which leads in particular to another proof of the regularity theorem.

Let Ω be an open set in R^n and let $P : C^\infty(\Omega) \to C^\infty(\Omega)$ be a linear differential operator of order m. If $u \in C_0^\infty(\Omega)$, we can write

$$(Pu)(x) = \int_{R^n} p(x,\xi)\,\hat{u}(\xi)\,e^{i\langle x,\xi\rangle}\,d\xi\,,$$

where $\hat{u}$ is the Fourier transform of u, *and* $p(x,\xi)$ *is a polynomial of degree* m *in* ξ.

We extend the class of differential operators by allowing more inclusive classes of functions p than the one above. Several such classes have been considered, and each of them is useful in some contexts. For the study of elliptic differential operators, the most suitable is the following.

Let m be a real number and let Ω be open in R^n. We define $S^m(\Omega)$ to be the set of C^∞ functions $p : \Omega \times R^n \to C$ with the following property. If $K \subset \Omega$ is compact, and if $\alpha,\beta \in (Z^+)^n$, then there exists $M>0$ such that

$$|D_\xi^\alpha D_x^\beta p(x,\xi)| \leq M \cdot (1+|\xi|)^{m-|\alpha|} \quad \text{for } x \in K, \xi \in R^n.$$

The operators from $C_0^\infty(\Omega)$ to $C^\infty(\Omega)$ defined by

$$(*) \qquad (Pu)(x) = \int_{R^n} p(x,\xi)\,\hat{u}(\xi)\,e^{i\langle x,\xi\rangle}\,d\xi\,,$$

when $p \in S^m(\Omega)$ are called pseudo-differential operators of order $\leqq m$; we denote the set of these operators by $L^m(\Omega)$. The function p is called the total symbol of the operator P.

The principal symbol of a pseudo-differential operator has to be defined differently from that of a differential operator. It is defined to be, simply, the residue class of the total symbol in the quotient space

$$S^m(\Omega)/S^{m-1}(\Omega).$$

One proves that the composite of two pseudo-differential operators P, Q of orders m,n respectively is a pseudo-differential operator of order $m+n$, and that the principal symbol of $P \circ Q$ is the residue class in $S^{m+n}(\Omega)/S^{m+n-1}(\Omega)$, of the product of the total symbols of P and Q. Moreover, if we identify $\Omega \times R^n$ in the obvious way with the cotangent bundle $T^*(\Omega)$ of Ω, then the isomorphism

$$S^m(\Omega)/S^{m-1}(\Omega) \to L^m(\Omega)/L^{m-1}(\Omega)$$

commutes with diffeomorphisms of Ω. All this enables one to define the space $L^m(E,F)$ of pseudo-differential operators of order m between complex vector bundles E and F on a manifold V. The principal symbol defines, as in the case of differential operators, a linear map $\sigma(\xi): E_x \to F_x$, where $\xi \in T^*_x(V)$.

Once this calculus of pseudo-differential operators has been set up, it is not difficult to prove the existence of a so-called parametrix for an elliptic pseudo-differential operator. More precisely, let E,F be complex vector bundles of the same rank on a paracompact manifold V. Let $P \in L^m(E,F)$, and suppose that P is *elliptic*; i.e. for any $\xi \neq 0$ in $T^*_x(V)$, the map $\sigma(\xi)$ is an isomorphism of E_x onto F_x. Then, there exists an operator Q in $L^{-m}(F,E)$ such that $Q \circ P - \mathrm{Id}_E$ and $P \circ Q - \mathrm{Id}_F$ are pseudo-differential operators of order $-\infty$, (i.e. are of order $\leqq -N$ for every $N > 0$). It is a simple matter to deduce the main results (Friedrichs' inequality, the regularity theorem and so on) on elliptic differential operators from this theorem.

For an outline of this theory, and generalisations to "Fourier integral operators", see Hörmander [11, 12]. The calculus of pseudo-differential operators is set out very elegantly in Malgrange [14]; see also Kohn–Nirenberg [13], Nirenberg [15] and Seeley [18].

It should be mentioned that the theory of pseudo-differential operators evolved from the work of Calderón and Zygmund on singular integrals; see [5], also E. M. Stein [20]. The calculus of pseudo-differential operators is also particularly well suited to the requirements of the Atiyah–Singer theorem [1], a deep result relating the topology of smooth manifolds to analysis on them.

There is, however, yet another reason for looking at the integral defining pseudo-differential operators, namely that it leads naturally to a subject of great contemporary interest. It is clear that the integral in (*) ignores, to a large extent, the fact that, for a fixed value of x, the function $p(x,\xi)$ might behave very differently in different regions of values of ξ. The only case in which this loss of information is not too significant is that of elliptic operators. It would therefore be natural to investigate the behaviour of the integral in (*) extended not over all R^n, but over small open sets in R^n. In the language of manifolds and their cotangent bundles, this amounts to considering local behaviour *in the cotangent bundle*. This leads to the very important theory of microlocalisation and microlocal operators. The following works will give the reader some idea of the power and richness of this subject: J. E. Björk [2], Guillemin, Kashiwara and Kawai [8], C. Fefferman and D. H. Phong [6], F. Pham [16].

References

[1] M. F. Atiyah and I. M. Singer, The index of elliptic operators I, III, *Annals of Math.* 87 (1968) 484–530, 546–604.

[2] J. E. Björk, *Rings of Differential Operators.* North-Holland, Amsterdam, 1979.

[3] E. Cartan, *Leçons sur les Invariants Intégraux*, Hermann, Paris, 1922.

[4] E. Cartan, *Leçons sur la Géométrie des Espaces de Riemann*, Gauthier-Villars, Paris, 1928.

[5] A. Calderón and A. Zygmund, (i) On the existence of certain singular integrals, *Acta Math.* 88 (1952) 85–139.
(ii) Singular integral operators and differential equations, *Amer. J. Math.* 79 (1957) 801–821.

[6] C. Fefferman and D. H. Phong, Symplectic geometry and positivity of pseudo-differential operators, *Proc. Nat. Acad. Sci. U.S.A.* 79 (1982) 710–713.

[7] G.B. Folland and J. J. Kohn, *The Neumann Problem for the Cauchy-Riemann Complex*, Annals of Math. Studies, Princeton, 1972.

[8] V. W. Guillemin, M. Kashiwara and T. Kawai, *Seminar on Microlocal Analysis*, Annals of Math. Studies, Princeton, 1979.

[9] L. Hörmander. *An Introduction to Complex Analysis in Several Variables*, North-Holland, Amsterdam, 1973.

[10] L. Hörmander, L^2 estimates and existence theorems for the $\bar{\partial}$ operator, *Acta Math.* 113 (1965) 89–152.

[11] L. Hörmander, The calculus of Fourier integral operators. pp. 33–57 in *Prospects in Mathematics*, Annals of Math. Studies, Princeton, 1971.

[12] L. Hörmander, Fourier integral operators I, *Acta Math.* 127 (1971) 79–183.

[13] J. J. Kohn and L. Nirenberg, On the algebra of pseudodifferential operators, *Comm. Pure Appl. Math.* 18 (1965) 269–305.

[14] B. Malgrange, *Opérateurs Pseudodifférentiels*, Univ. de Genève, circa 1968.

[15] L. Nirenberg, Pseudodifferential operators, in: *Global Analysis*, Proceedings of Symposia in Pure Math. Amer. Math. Soc. Vol. 16 (1970) 149–167.

[16] F. Pham, *Singularités des Systèmes Différentiels de Gauss–Manin*, Birkhäuser, Boston, 1979.

[17] H. Poincaré, Analysis Situs, *École Polytéchnique* 1 (1895) 1–121. {*Oevres* Vol. VI 193–288}.

[18] R. T. Seeley, Integro-differential operators on vector bundles, *Trans. Amer. Math. Soc.* 117 (1965) 167–204.

[19] Y. T. Siu and S. T. Yau. Complete Kähler manifolds with nonpositive curvature of faster than quadratic decay. *Annals of Math.* 105 (1977) 225–264.

[20] E. M. Stein, *Singular Integrals and Differentiability Properties of Functions*, Princeton Univ. Press, Princeton, 1970.

[21] V. Volterra, (i) Delle variabili complesse negli iperspazi, *Rend. Accad. dei Lincei* 5 (1889) 158–165; 291–299. {*Opere* Vol. 1, 403–410; 411–419}.
(ii) Sulle funzioni coniugate, *Rend. Accad. dei Lincei* 5 (1889) 599–611. {*Opere* Vol. 1, 420–432}.

Contents

CHAPTER 1

Differentiable functions in R^n

Notation. We shall use the following notation. We use R, C, Q, Z to denote, respectively, the field of real numbers, the field of complex numbers, the field of rational numbers and the ring of integers. We shall look upon the first two as being provided with their usual topology. R^n, C^n, ... will denote the Cartesian product of R, C, ..., respectively, so that, e.g.,

$$R^n = \{(x_1, \ldots, x_n) | \, x_j \in R, j = 1, \ldots, n\}.$$

The notations R^+, Q^+, Z^+ stand for the sets of non-negative elements of R, Q, Z respectively.

For the most part, α, β stand for n-tuples of non-negative integers, $\alpha = (\alpha_1, \ldots, \alpha_n)$, $\beta = (\beta_1, \ldots, \beta_n)$, $\alpha_j, \beta_j \in Z^+$. We then set

$$|\alpha| = \alpha_1 + \ldots + \alpha_n, \qquad \alpha! = \alpha_1! \cdots \alpha_n!,$$

$$\binom{\alpha}{\beta} = \frac{\alpha!}{\beta!(\alpha-\beta)!} \qquad \text{if} \quad \beta_j \leqq \alpha_j.$$

We write $\beta \leqq \alpha$ if $\beta_j \leqq \alpha_j$ and $\beta < \alpha$ if $\beta \leqq \alpha$ and $\beta \neq \alpha$.

We denote a point of R^n [C^n] by $x = (x_1, \ldots, x_n)$ [$z = (z_1, \ldots, z_n)$]. Then

$$|x| = \max_j |x_j|, \qquad |z| = \max_j |z_j|,$$

$$||x|| = (|x_1|^2 + \ldots + |x_n|^2)^{\frac{1}{2}}, \quad ||z|| = (|z_1|^2 + \ldots + |z_n|^2)^{\frac{1}{2}}$$

and

$$x^\alpha = x_1^{\alpha_1} \cdots x_n^{\alpha_n}, \qquad z^\alpha = z_1^{\alpha_1} \cdots z_n^{\alpha_n}.$$

If X is a (Hausdorff) topological space and S a subset of X, we denote by $\mathring{S}$ the interior of S, i.e., the largest open set contained in S. If S_1, S_2 are two subsets of X, we write $S_1 \Subset S_2$ if S_1 is relatively compact in S_2; i.e., if the closure of S_1 in S_2 is compact.

If f is a map of an open set Ω in R^n into R^q and λ a function $\geqq 0$ in Ω, we write

$$f(x) = O(\lambda(x)), \qquad [\text{or } f = O(\lambda)]$$

if there is a constant $C > 0$ such that $|f(x)| \leqq C\lambda(x)$ for all $x \in \Omega$. In addition, if $a \in \Omega$, we write

$$f(x) = o(\lambda(x)),$$

as $x \to a$ (or $|x - a| \to 0$), if there is a map ε: $\Omega \to R^+$ such that $\varepsilon(x) \to 0$ as $x \to a$ and $|f(x)| \leqq \varepsilon(x)\lambda(x)$.

Similar notation is used when a is replaced by a "point at infinity".

§ 1.1 Taylor's formula

Let Ω be an open set in R^n and k an integer $\geqq 0$. We denote by $C^k(\Omega)$ the set of real-valued functions f on Ω which possess continuous partial derivatives of order $\leqq k$, i.e., for which the derivatives

$$\frac{\partial^{\alpha_1 + \dots + \alpha_n} f}{\partial x_1^{\alpha_1} \cdots \partial x_n^{\alpha_n}}$$

exist and are continuous on Ω for $|\alpha| = \alpha_1 + \dots + \alpha_n \leqq k$. We denote by $C^\infty(\Omega)$ the set of functions belonging to $C^k(\Omega)$ for all $k \geqq 0$. Functions in $C^k(\Omega)$ are called C^k functions on Ω. For $f \in C^k(\Omega)$ and $|\alpha| \leqq k$, we denote the partial derivative

$$\frac{\partial^{\alpha_1 + \dots + \alpha_n} f}{\partial x_1^{\alpha_1} \cdots \partial x_n^{\alpha_n}}$$

by

$$D^\alpha f = \left(\frac{\partial}{\partial x_1}\right)^{\alpha_1} \cdots \left(\frac{\partial}{\partial x_n}\right)^{\alpha_n} f.$$

The order in which the differentiations are performed is irrelevant.

For any function f (not necessarily continuous) defined on Ω, we denote by supp(f) the closure in Ω of the set

$$\{x \in \Omega \mid f(x) \neq 0\}.$$

Supp(f) is called the *support* of f. $C_0^\infty(\Omega)$ stands for the set of $f \in C^k(\Omega)$ such that supp(f) is compact.

If E is a finite-dimensional R vector space, we denote by $C^k(\Omega, E)$, $C_0^k(\Omega, E), \ldots$ the set of all mappings $f: \Omega \to E$ such that, for any (continuous) linear functional l on E, the function $l \circ f \in C^k(\Omega)$, $C_0^k(\Omega), \ldots$.

If $e_1, \ldots, e_q$ is an R basis of E, and $f: \Omega \to E$ is a map, for each $x \in \Omega$, there are real numbers $f_1(x), \ldots, f_q(x)$ such that

$$f(x) = \sum_{j=1}^{q} f_j(x) e_j .$$

It is easily checked that

$$f \in C^k(\Omega, E), C_0^k(\Omega, E), \ldots$$

if and only if

$$f_j \in C^k(\Omega), C_0^k(\Omega), \ldots \qquad \text{for } j = 1, \ldots, q.$$

Elements of $C^k(\Omega, E)$ are called C^k mappings of Ω into E. If $E = R^q$ we write $C^k(\Omega, q)$, $C_0^k(\Omega, q), \ldots$ for $C^k(\Omega, E)$, $C_0^k(\Omega, E), \ldots$. For $f \in C^k(\Omega, E)$, we can define the derivatives $D^\alpha f$ for $|\alpha| \leqq k$. Then

$$D^\alpha f \in C^{k-|\alpha|}(\Omega, E).$$

We shall identify $f \in C_0^k(\Omega)$ with the element $g \in C_0^k(R^n)$ which $= f$ on Ω and $= 0$ on $R^n - \Omega$.

We shall often deal with complex-valued functions on Ω (or mappings of Ω into C^q). We shall then use the notation $C^k(\Omega)$, $C^k(\Omega, q), \ldots$ for $C^k(\Omega, C)$, $C^k(\Omega, C^q), \ldots$ if no confusion is likely.

A real-valued function f defined on Ω is called (*real*) *analytic* if, for any $a = (a_1, \ldots, a_n) \in \Omega$, there exists a power series

$$P_a(x) \equiv \sum_\alpha c_\alpha (x-a)^\alpha = \sum_{\alpha_j \geqq 0} c_{\alpha_1 \ldots \alpha_n} (x_1 - a_1)^{\alpha_1} \ldots (x_n - a_n)^{\alpha_n},$$

which converges to $f(x)$ for x in a neighbourhood U of a. The series then converges uniformly to f on compact subsets of U (so that f

is continuous) and so does the differentiated series. Hence $f \in C^\infty(\Omega)$ and for any $\beta = (\beta_1, \ldots, \beta_n)$,

$$D^\beta f(x) = D^\beta P_a(x) = \sum_\alpha c_\alpha D^\beta (x-a)^\alpha.$$

Moreover, the series is uniquely determined by f; in fact

$$c_\alpha = \frac{1}{\alpha!} D^\alpha f(a).$$

Analytic maps of Ω into a finite dimensional vector space are defined in the same way as above.

If U, V are open sets in R^n and $f\colon U \to V$ is a homeomorphism such that both f and f^{-1} are C^k mappings, we say that f is a C^k *diffeomorphism* (or just diffeomorphism) of U onto V. If $U = V$, we call f a C^k *automorphism*.

If f and f^{-1} are real analytic, we speak of an analytic isomorphism (or automorphism if $U = V$).

If U is an open set in C^n and f a complex-valued function on U, f is called *holomorphic* if for any $a \in U$, there is a power series $\sum c_\alpha (z-a)^\alpha$, which converges to $f(z)$ for all z in a neighbourhood of a.

If E is a finite dimensional C vector space, a map $f\colon U \to E$ is called *holomorphic* if for any C linear function l on E, $l \circ f$ is holomorphic. A map $f\colon \Omega \to C^q$ is holomorphic if and only if, when we write $f = (f_1, \ldots, f_q)$, each f_j is a holomorphic function.

A map $f\colon U \to V$ (open sets in C^n) is called a *C analytic isomorphism* (or, by abuse of language, an analytic isomorphism if no confusion is likely) if f and f^{-1} are holomorphic. A theorem of Osgood (*see* e.g. HERVÉ [1963]), which we shall not prove in this book, asserts that a one-one holomorphic map of U onto V is a C analytic isomorphism. There is no analogue for C^k or real analytic maps.

We shall assume some elementary properties of holomorphic functions. These are proved in most books on several complex variables, *see* e.g. HERVÉ [1963], and HÖRMANDER [1966].

1.1.1 CAUCHY-RIEMANN EQUATIONS. A function defined on an open set $U \subset C^n$ is holomorphic if and only if it is continuous and, for any j, $1 \leqq j \leqq n$, the partial derivatives

$$\frac{\partial f}{\partial \bar{z}_j} \equiv \frac{1}{2}\left(\frac{\partial f}{\partial x_j} + \mathrm{i}\frac{\partial f}{\partial y_j}\right)$$

exist and are 0. Here $z_j = x_j + \mathrm{i}y_j$, x_j, y_j are real and $\mathrm{i} = \sqrt{-1}$.

We also set

$$\frac{\partial f}{\partial z_j} = \frac{1}{2}\left(\frac{\partial f}{\partial x_j} - \mathrm{i}\frac{\partial f}{\partial y_j}\right).$$

For a *holomorphic* function f on U, we write

$$D^\alpha f = \left(\frac{\partial}{\partial z_1}\right)^{\alpha_1} \cdots \left(\frac{\partial}{\partial z_n}\right)^{\alpha_n} f.$$

In view of the equations 1.1.1, we have

$$D^\alpha f = \left(\frac{\partial}{\partial x_1}\right)^{\alpha_1} \cdots \left(\frac{\partial}{\partial x_n}\right)^{\alpha_n} f.$$

A basic theorem of Hartogs (*see* HÖRMANDER [1966]) asserts that the condition of continuity is superfluous in the Cauchy-Riemann-equations 1.1.1; we shall not prove this here.

1.1.2 PRINCIPLE OF ANALYTIC CONTINUATION. If f is holomorphic (real analytic) in a connected open set $U(\Omega)$ in C^n (R^n) and $D^\alpha f(a) = 0$ for all $\alpha = (\alpha_1, \ldots, \alpha_n)$ and some $a \in U(\Omega)$, then $f \equiv 0$. In particular, if f vanishes on a non-empty open subset of $U(\Omega)$, then $f \equiv 0$.

1.1.3 WEIERSTRASS' THEOREM. If $\{f_\nu\}$ is a sequence of holomorphic functions, converging uniformly on compact subsets of U to a function f, then f is holomorphic in U. Moreover, for any α, $\{D^\alpha f_\nu\}$ converges to $D^\alpha f$, uniformly on compact sets.

1.1.3′ MONTEL'S THEOREM. If $\mathfrak{S} = \{f\}$ is a family of holomorphic functions in U which is uniformly bounded on compact subsets K of U:

$$|f(x)| \leqq M \qquad \text{for all } x \in K,\ f \in \mathfrak{S},$$

then any sequence of elements of $\mathfrak{S}$ contains a subsequence which converges uniformly on compact subsets of U.

1.1.3″ THE MAXIMUM PRINCIPLE. Let f be holomorphic in a connected open set U in C^n. Then, the map $f: U \to C$ is either constant or open. In particular, if U is bounded and we set

$$M = \sup_{\zeta \in \partial U} \overline{\lim_{z \to \zeta,\, z \in U}} |f(z)|,$$

we have $|f(z)| < M$ for all $z \in U$ unless f is constant.

1.1.4 CAUCHY'S INEQUALITIES. If f is holomorphic in U and $|f(z)| \leqq M$ for $z \in U$, then for any compact set $K \subset U$ and any α, we have

$$|D^\alpha f(z)| \leqq M\, \alpha!\, \delta^{-|\alpha|} \qquad \text{for } z \in K,$$

where δ is the distance of K from the boundary of U.

1.1.5 LEMMA. Let f be real analytic in $\Omega \subset R^n$. We look upon R^n as a closed subset of C^n. Then there exists an open set $U \subset C^n$, $U \cap R^n = \Omega$ and a holomorphic function F in U with $F|\Omega = f$.

PROOF. Let $a \in \Omega$ and let $P_a(x) = \sum c_\alpha (x-a)^\alpha$ be a power series converging to $f(x)$ for $|x-a| < r_a$, $r_a > 0$. Define

$$U_a = \{z \in C^n |\, |z-a| < r_a\}.$$

Then, for $z \in U_a$, $P_a(z) = \sum c_\alpha (z-a)^\alpha$ converges and is a holomorphic function on U_a.

Let $U = \bigcup_{a \in \Omega} U_a$. We assert that if $U_a \cap U_b = U_{a,b} \neq \emptyset$, then $P_a = P_b$ in $U_{a,b}$. In fact, $U_{a,b}$ is convex, hence connected. Further, if $U_{a,b} \neq \emptyset$, then $U_{a,b} \cap R^n \neq \emptyset$ and, for any $c \in U_{a,b} \cap R^n$, we have

$$D^\alpha P_a(c) = D^\alpha f(c) = D^\alpha P_b(c),$$

and we may apply principle 1.1.2. Hence we may define a holomorphic function F on U by setting $F|U_a = P_a$. Clearly $F|\Omega = f$.

We return now to real valued functions. Let N be a neighbourhood of the closed unit interval $0 \leqq t \leqq 1$ in R, and let $f \in C^k(N)$, $k \geqq 1$. Then we have:

1.1.6 LEMMA. There is a ξ with $0 \leqq \xi \leqq 1$ such that

$$f(1) = \sum_{\nu=0}^{k-1} \frac{f^{(\nu)}(0)}{\nu!} + \frac{f^{(k)}(\xi)}{k!},$$

where

$$f^{(\nu)}(t) = \left(\frac{\mathrm{d}}{\mathrm{d}t}\right)^{\nu} f(t).$$

PROOF. For a continuous function g, on N, set

$$I_0(g, t) = g(t), \qquad I_r(g, t) = \int_0^t I_{r-1}(g, s)\, \mathrm{d}s, \qquad r \geqq 1.$$

Clearly, if $g \in C^k(N)$ and $g^{(\nu)}(0) = 0$ for $0 \leqq \nu \leqq k-1$, we have

$$g(t) = I_k(g^{(k)}, t).$$

If we apply this to

$$g(t) = f(t) - \sum_{\nu=0}^{k-1} \frac{f^{(\nu)}(0)}{\nu!} t^{\nu},$$

we obtain

1.1.7 $$f(1) - \sum_{\nu=0}^{k-1} \frac{f^{(\nu)}(0)}{\nu!} = I_k(g^{(k)}, 1) = I_k(f^{(k)}, 1).$$

If m and M denote respectively the lower and upper bounds of $f^{(k)}$ in $[0, 1]$, we have, clearly,

$$\frac{m}{k!} \leqq I_k(f^{(k)}, 1) \leqq \frac{M}{k!}.$$

Since $f^{(k)}$ is continuous, and so takes all values between m and M, there is a ξ, $0 \leqq \xi \leqq 1$ for which

$$I_k(f^{(k)}, 1) = \frac{1}{k!} f^{(k)}(\xi).$$

This proves the lemma.

It is easy to prove by induction that

$$I_k(g,t) = \frac{1}{(k-1)!} \int_0^t g(s)(t-s)^{k-1}\, \mathrm{d}s.$$

Hence, (1.1.7) can be written in the form:

1.1.8 $$f(1) = \sum_{\nu=0}^{k-1} \frac{f^{(\nu)}(0)}{\nu!} + \frac{1}{(k-1)!}\int_0^1 (1-t)^{k-1} f^{(k)}(t)\,dt.$$

1.1.9 THEOREM (TAYLOR'S FORMULA). Let Ω be an open set in R^n and let $f \in C^k(\Omega)$. Let $x, y \in \Omega$ and suppose that the closed line segment $[x, y]$ joining x and y is contained in Ω. Then we have

$$f(x) = \sum_{|\alpha| \leqq k-1} \frac{1}{\alpha!} D^\alpha f(y)(x-y)^\alpha + \sum_{|\alpha|=k} \frac{1}{\alpha!} D^\alpha f(\xi)(x-y)^\alpha,$$

where ξ is a point of $[x, y]$.

PROOF. This follows at once from lemma 1.1.6 applied to the function

$$g(t) = f(y+t(x-y)),$$

which belongs to $C^k(N)$ for a suitable neighbourhood N of $[0, 1]$.

If $f \in C^m(\Omega)$, and S is a subset of Ω, we set

$$\|f\|_m^S = \sum_{|\alpha| \leqq m} \frac{1}{\alpha!} \sup_{x \in S} |D^\alpha f(x)|.$$

Note that if $f, g \in C^m(\Omega)$, then we have

1.1.10 $$\|fg\|_m^S \leqq \|f\|_m^S \, \|g\|_m^S, \qquad \|f+g\|_m^S \leqq \|f\|_m^S + \|g\|_m^S.$$

We define a topology on $C^k(\Omega)$, k finite, as follows. A fundamental system of neighbourhoods of $g \in C^k(\Omega)$ is given by the sets

$$B(g,K,\varepsilon) = \{f \in C^k(\Omega) |\ \|f-g\|_k^K < \varepsilon\};$$

here ε runs over all positive real numbers and K over all compact subsets of Ω. The corresponding topology on $C^\infty(\Omega)$ is obtained by taking for a fundamental system of neighbourhoods the sets

$$\{f \in C^\infty(\Omega) |\ \|f-g\|_m^K < \varepsilon\},$$

where ε, K are as before, and m is an arbitrary positive integer.

It is clear that one may introduce similar topologies on $C^k(\Omega,E)$

($0 \leqq k \leqq \infty$) for a finite dimensional vector space E (of dimension q) which makes this space isomorphic algebraically and topologically, to the Cartesian product $(C^k(\Omega))^q$. This isomorphism depends on a choice of basis for E. However, $C^k(\Omega, E)$ is canonically isomorphic with $C^k(\Omega) \otimes E$.

It is easily seen that the above topology on $C^k(\Omega)$ has a countable base. A sequence $\{f_\nu\}$ converges to 0, if and only if $D^\alpha f_\nu \to 0$ uniformly on compact sets for all α with $|\alpha| \leqq k$ (all α if $k = \infty$). Further the above topology is metrisable; if $k < \infty$ one may take, as metric the function

$$d(f, g) = \sum_{\nu=0}^{\infty} 2^{-\nu} \frac{\|f-g\|_k^{K_\nu}}{1+\|f-g\|_k^{K_\nu}},$$

where $\{K_\nu\}$ is a sequence of compact sets with $K_\nu \subset \mathring{K}_{\nu+1}$ (interior of $K_{\nu+1}$) and $\bigcup K_\nu = \Omega$. If $k = \infty$, one may take instead the function

$$\sum_{\nu=0}^{\infty} 2^{-\nu} \frac{\|f-g\|_\nu^{K_\nu}}{1+\|f-g\|_\nu^{K_\nu}}.$$

1.1.11 THEOREM. For $0 \leqq k \leqq \infty$, $C^k(\Omega)$ is a complete metric space.

PROOF. We have only to show, that if $\{g_\nu\}$ is a sequence of functions in $C^k(\Omega)$ and

$$\|g_\nu - g_\mu\|_m^K \to 0$$

as $\mu, \nu \to \infty$ for all integers $m \leqq k$ and all compact $K \subset \Omega$ (the condition $m \leqq \infty$ is empty), then there is $g \in C^k(\Omega)$ such that

$$\|g_\nu - g\|_m^K \to 0,$$

for $m \leqq k$ and all compact K. Now, for any α, $|\alpha| \leqq k$, there is a continuous function g_α such that

$$\|D^\alpha(g_\nu) - g_\alpha\|_0^K \to 0 \quad \text{as} \quad \nu \to \infty$$

(since clearly $D^\alpha(g_\nu - g_\mu) \to 0$ uniformly on compact sets). We have only to prove that $g = g_0 \in C^k(\Omega)$ and $D^\alpha g = g_\alpha$. To do this, it is sufficient to show that if $|\alpha| \leqq k-1$, and $\beta = (\beta_1, \ldots, \beta_n)$ is such that $|\beta| = 1$, then $g_\alpha \in C^1(\Omega)$ and $D^\beta g_\alpha = g_{\alpha+\beta}$.

If $a \in \Omega$ and x is near a, we have, by Taylor's formula:

1.1.12 $$D^\alpha g_\nu(x) - D^\alpha g_\nu(a) = \sum_{|\beta|=1} D^{\alpha+\beta} g_\nu(\xi_\nu)(x-a)^\beta,$$

where ξ_ν is a point on the segment $[a, x]$. We can choose a subsequence $\{\nu_p\}$ such that

$$\xi_{\nu_p} \to \xi \in [a,x].$$

If we replace ν by ν_p in (1.1.12) and let $p \to \infty$, we obtain

$$\begin{aligned} g_\alpha(x) - g_\alpha(a) &= \sum_{|\beta|=1} g_{\alpha+\beta}(\xi)(x-a)^\beta \\ &= \sum_{|\beta|=1} g_{\alpha+\beta}(a)(x-a)^\beta + \mathrm{o}(|x-a|), \end{aligned}$$

where $\mathrm{o}(|x-a|)$ is a function which tends to zero faster than $|x-a|$ as $x \to a$. The latter equality follows from the continuity of $g_{\alpha+\beta}$. This implies that $g_\alpha \in C^1(\Omega)$ and that for $|\beta| = 1$, we have $D^\beta g_\alpha = g_{\alpha+\beta}$.

1.1.13 REMARK. The same result is clearly also true for $C^k(\Omega, E)$. Another consequence of Taylor's formula is the following:

1.1.14 PROPOSITION. *If $f \in C^\infty(\Omega)$, then f is analytic if and only if for any compact set $K \subset \Omega$, there exists an $M > 0$ such that*

1.1.15 $$|D^\alpha f(x)| \leq M^{|\alpha|+1}\alpha! \qquad \text{for } x \in K \text{ and all } \alpha.$$

PROOF. If f is analytic the inequalities follow from lemma 1.1.5 and Cauchy's inequalities 1.1.4. Conversely, if (1.1.15) holds and x belongs to a compact convex neighbourhood K of $a \in \Omega$ and for $\xi \in [a, x]$, we have

$$\Big|\sum_{|\alpha|=k} \frac{1}{\alpha!} D^\alpha f(\xi)(x-a)^\alpha\Big| \leq k^n M^{k+1}|x-a|^k.$$

If $|x-a| < M^{-1}$, Taylor's formula implies that

$$\sum \frac{1}{\alpha!} D^\alpha f(a)(x-a)^\alpha$$

converges to $f(x)$.

1.1.16 REMARK. It is easily verified that (1.1.15) is equivalent to the existence of $M' > 0$ such that

$$|D^\alpha f(x)| \leqq M'^{|\alpha|+1}(|\alpha|)! \qquad \text{for} \quad x \in K \text{ and all } \alpha.$$

§ 1.2 Partitions of unity

Before we state the main result, we introduce a few definitions.

A family $\{E_i\}_{i \in I}$ of subsets of a topological space X is called *locally finite* if every point $a \in X$ has a neighbourhood U such that

$$\{i \in I |\ E_i \cap U \neq \emptyset\} \text{ is finite.}$$

A family $\{E'_j\}_{j \in J}$ is called *refinement* of a family $\{E_i\}_{i \in I}$ if there is a mapping

$$r: J \to I,$$

(called the refinement map) such that $E'_j \subset E_{r(j)}$ for all $j \in J$.

We shall use the following proposition, due to DIEUDONNÉ [1944]. Proofs are to be found in the standard texts on set topology (e.g. BOURBAKI [1965, 1958].

1.2.1 PROPOSITION. If X is a locally compact Hausdorff space, which is a countable union of compact sets, then X is paracompact, i.e., any open covering has a locally finite refinement, which is also an open covering. Further, for any locally finite covering $\{U_i\}_{i \in I}$ of X, there is an open covering $\{V_i\}_{i \in I}$ of X (with the same indexing set) such that $\bar{V}_i \subset U_i$.

1.2.2 DEFINITION. Let Ω be an open set in R^n and $\{U_i\}_{i \in I}$ an open covering. A family $\{\varphi_i\}_{i \in I}$ of C^∞ functions is called a partition of unity, subordinate to the covering $\{U_i\}_{i \in I}$, if

$$0 \leqq \varphi_i \leqq 1; \qquad \operatorname{supp}(\varphi_i) \subset U_i;$$

the family $\{\operatorname{supp}(\varphi_i)\}$ is locally finite and

$$\sum_{i \in I} \varphi_i(x) = 1, \text{ for any } x \in \Omega.$$

1.2.3 THEOREM. Let Ω be an open set in R^n and $\{U_i\}_{i \in I}$ an open covering. Then there exists a partition of unity subordinate to $\{U_i\}$.

We need two lemmas for the proof.

1.2.4 LEMMA. There exists a C^∞ function η on R^n with $\eta \geqq 0, \eta(0) > 0$, $\operatorname{supp}(\eta) \subset \{x \mid \|x\| < 1\}$.

PROOF. Let $0 < c < 1$ and s be the C^∞ function on R^1 defined by

$$s(r) = \begin{cases} \exp(-1/(c-r)) & \text{if } r < c; \\ 0 & \text{if } r \geqq c. \end{cases}$$

We may take

$$\eta(x) = s(x_1^2 + \ldots + x_n^2).$$

1.2.5 LEMMA. Let K be a compact set in R^n and U be an open set containing K. Then there is a C^∞ function φ on R^n with $\varphi(x) \geqq 0$ for all x, $\varphi(x) > 0$ for $x \in K$ and $\operatorname{supp}(\varphi) \subset U$.

PROOF. Let δ be the distance of K from $R^n - U$ if $U \neq R^n$, $\delta = 1$ if $U = R^n$. For $a \in K$, let

$$\varphi_a(x) = \eta\left(\frac{x-a}{\delta}\right),$$

where η is as in lemma 1.2.4. Let

$$V_a = \{x \in R^n \mid \varphi_a(x) > 0\}.$$

Then $a \in V_a \subset U$. Since K is compact, there are finitely many points $a_1, \ldots, a_p \in K$ for which

$$K \subset V_{a_1} \cup \ldots \cup V_{a_p}.$$

We may take $\varphi = \sum_{j=1}^p \varphi_{a_j}$.

PROOF OF THEOREM 1.2.3. Let $\{V_j\}_{j \in J}$ be a locally finite refinement of $\{U_i\}_{i \in I}$ by relatively compact open subsets of Ω (which exists by proposition 1.2.1). Let $\{W_j\}_{j \in J}$ be an open covering of Ω such that $\overline{W}_j \subset V_j$ (prop. 1.2.1). By lemma 1.2.5, there exists a C^∞ function ψ_j on Ω, $\psi_j(x) > 0$ for $x \in \overline{W}_j$ and $\operatorname{supp}(\psi_j) \subset V_j$, $0 \leqq \psi_j$. Let

$$\varphi_j' = \frac{\psi_j}{\sum_{j' \in J} \psi_{j'}}$$

[Since $\{V_j\}$ is locally finite, $\sum_{j' \in J} \psi_{j'}$ is defined, belongs to $C^\infty(\Omega)$, and is everywhere > 0 since $\psi_j > 0$ on W_j and $\bigcup W_j = \Omega$.] Clearly

$$\varphi'_j \geqq 0, \qquad \operatorname{supp}(\varphi'_j) \subset V_j \quad \text{and} \quad \sum_{j \in J} \varphi'_j = 1.$$

Let $r: J \to I$ be a map such that $V_j \subset U_{r(j)}$. Let $J_i \subset J$ be the set $r^{-1}(i)$. Let $\varphi_i = \sum_{j \in J_i} \varphi'_j$, where an empty sum stands for 0. Since the sets J_i are mutually disjoint and cover J. we have

$$\sum_{i \in I} \varphi_i = \sum_{j \in J} \varphi'_j = 1.$$

Clearly $\operatorname{supp}(\varphi_i) \subset U_i$. Since the family $\{\operatorname{supp}(\varphi'_j)\}$ is locally finite, so is $\{\operatorname{supp}(\varphi_i)\}$.

1.2.6 COROLLARY. Let Ω be open in R^n, X a closed subset of Ω and U an open subset of Ω containing X. Then, there exists a C^∞ function ψ on Ω such that $\psi(x) = 1$ if $x \in X$, $\psi(x) = 0$ if $x \in \Omega - U$ and $0 \leqq \psi \leqq 1$ everywhere.

PROOF. By theorem 1.2.3, there exist C^∞ functions $\varphi_1, \varphi_2 \geqq 0$ with $\operatorname{supp}(\varphi_1) \subset U$, $\operatorname{supp}(\varphi_2) \subset \Omega - X$ and $\varphi_1 + \varphi_2 = 1$ on Ω. We may take $\psi = \varphi_1$.

1.2.7 LEMMA. If $\{U_i\}_{i \in I}$ is an open covering of Ω, there exist C^∞ functions ψ_i with $\operatorname{supp}(\psi_i) \subset U_i$, $0 \leqq \psi_i \leqq 1$ and $\sum_i \psi_i^2 = 1$ on Ω.

PROOF. If $\{\varphi_i\}$ is a partition of unity subordinate to $\{U_i\}$, we may take

$$\psi_i = \frac{\varphi_i}{(\sum_{j \in I} \varphi_j^2)^{\frac{1}{2}}}.$$

§ 1.3 Inverse functions, implicit functions and the rank theorem

Let Ω be an open set in R^n and let $f \in C^1(\Omega, m)$, i.e., f is a C^1 map of Ω into R^m. Let $a \in \Omega$.

1.3.1 DEFINITION. $(\mathrm{d}f)(a)$ is defined to be the R linear map of R^n into R^m for which

$$(\mathrm{d}f)(a)(v_1, \ldots, v_n) = (w_1, \ldots, w_m),$$

where

$$w_j = \sum_{k=1}^{n} \frac{\partial f_j}{\partial x_k}(a) v_k .$$

This map $(\mathrm{d}f)(a)$ is called the differential of f at a.

For a holomorphic map f of an open set in C^n into C^m, we define the differential $(\mathrm{d}f)(a)$, in the same way, i.e.,

$$(\mathrm{d}f)(a)(v_1, \ldots, v_n) = (w_1, \ldots, w_m), \qquad w_j = \sum_{k=1}^{n} \frac{\partial f_j}{\partial z_k}(a) v_k .$$

This is, of course, a C linear map of C^n into C^m. One can identify this with the map given above under the natural identification of C with R^2 using the Cauchy–Riemann equations.

1.3.2 THEOREM. If f is a C^1 map of Ω into R^n and, for a point $a \in \Omega$, $(\mathrm{d}f)(a)$ is an isomorphism of R^n onto itself, then there exist neighbourhoods U of a and V of $f(a)$ such that $f|U$ is a homeomorphism onto V.

PROOF. Without loss of generality, we may suppose that $a = 0$, $f(a) = 0$. Since $(\mathrm{d}f)(a) = A$ is an isomorphism of R^n onto itself, we can replace f by $A^{-1} \circ f$ and assume that $(\mathrm{d}f)(a)$ is the identity. Let g by defined on Ω by

$$g(x) = f(x) - x.$$

Then clearly $(\mathrm{d}g)(a) = 0$. This implies that there is a neighbourhood

$$W \text{ of } 0, \qquad \overline{W} \subset \Omega, \qquad W = \{x \mid |x_j| < r\},$$

such that $x, y \in \overline{W}$ implies that

$$|g(x) - g(y)| \leqq \tfrac{1}{2}|x - y|.$$

Clearly then, if x, $y \in W$, we have

$$|f(x) - f(y)| \geqq \tfrac{1}{2}|x - y|,$$

so that f is injective on W. Let $V = \{x \mid |x_j| < \frac{1}{2}r\}$ and $U = W \cap f^{-1}(V)$. Define $\varphi_0 \colon V \to W$ by $\varphi_0(y) = 0$ and, by induction,

$$\varphi_\nu(y) = y - g(\varphi_{\nu-1}(y)).$$

It is easily verified, by induction, that

$$\varphi_\nu(V) \subset W, \qquad \nu \geqq 0,$$

and further,

$$|\varphi_\nu(y)-\varphi_{\nu-1}(y)| = |g(\varphi_{\nu-1}(y))-g(\varphi_{\nu-2}(y))| \leqq r2^{-\nu} \qquad (\nu \geqq 2)$$

and this is true also for $\nu = 1$. Hence $\{\varphi_\nu\}$ converges uniformly to a map $\varphi: V \to R^n$. Since $\varphi_\nu(V) \subset W$, we have $\varphi(V) \subset \overline{W}$ and

$$\varphi(y) = y-g(\varphi(y)).$$

Since $|y| < r/2$ on V and $|g(\varphi(y))| \leqq r/2$, it follows that $\varphi(V) \subset W$; moreover, $f(\varphi(y)) = \varphi(y)+g(\varphi(y)) = y$. Since $f|W$ is injective, φ is the inverse of f. Clearly φ is continuous since the φ_ν are, which proves the theorem. We shall see later that $\varphi \in C^1(V, n)$.

1.3.3 REMARK. The theorem has an analogue for holomorphic maps. If Ω is an open set in C^n and $f: \Omega \to C^n$ a holomorphic map such that $(\mathrm{d}f)(a)$ is an isomorphism for some $a \in \Omega$, then there are neighbourhoods U of a and V of $f(a)$ such that $f|U$ is a homeomorphism of U onto V and *the inverse map of* $f|U$ *is holomorphic*. The proof is identical with the one given above. We define U, V and the φ_ν as above; $\{\varphi_\nu\}$ converges uniformly to the inverse φ of $f|U$. Since each φ_ν is holomorphic, so is φ by theorem 1.1.3.

1.3.4 DEFINITION. Let Ω_1, Ω_2 be open subsets of R^{n_1}, R^{n_2} respectively, let f be a C^1 map of $\Omega_1 \times \Omega_2$ into R^p and let $(a, b) \in \Omega_1 \times \Omega_2$. Define a map $g: \Omega_2 \to R^p$ by $g(y) = f(a, y)$. The partial differential $(\mathrm{d}_2 f)(a, b)$ is defined to be the linear map $(\mathrm{d}g)(b)$ of R^{n_2} into R^p. The partial differential $(\mathrm{d}_1 f)(a, b)$ is defined similarly.

1.3.5 THEOREM. Let Ω_1, Ω_2 be open sets in R^{n_1}, R^{n_2} respectively, let f be a C^1 map of $\Omega_1 \times \Omega_2$ into R^{n_2}. Suppose that for some $(a, b) \in \Omega_1 \times \Omega_2$, we have $f(a, b) = 0$, and rank $(\mathrm{d}_2 f)(a, b) = n_2$. Then there is a neighbourhood $U_1 \times U_2$ of (a, b) such that for any $x \in U_1$, there is a unique $y = y(x) \in U_2$ such that $f(x, y(x)) = 0$. The map $x \mapsto y(x)$ is continuous.

PROOF. Consider the map $F: \Omega_1 \times \Omega_2 \to R^{n_1+n_2}$ defined by

$$F(x, y) = (x, f(x, y)).$$

Then $(d_2f)(a, b)$ has rank n_2 if and only if $(dF)(a, b)$ is an isomorphism. Hence, by theorem 1.3.2, there is a neighbourhood $U \times U_2$ of (a, b) and a neighbourhood W of $(a, 0)$ such that $F|U \times U_2 \to W$ is a homeomorphism. Let $\varphi: W \to U \times U_2$ be the continuous inverse of $f|U \times U_2$. There is a neighbourhood U_1 of a such that $x \in U_1$ implies that $(x, 0) \in W$. For $x \in U_1$, let $y(x)$ be the projection of $\varphi(x, 0)$ on U_2. Clearly if $y \in U_2$ satisfies $f(x, y) = 0$, then $y = y(x)$. Further, $x \mapsto y(x)$ is a continuous map with $f(x, y(x)) = 0$.

1.3.6 REMARK. The theorem has an obvious analogue for holomorphic maps $f: \Omega_1 \times \Omega_2 \to C^{n_2}$ (with the obvious notation); $y(x)$ is then holomorphic.

1.3.7 LEMMA. With the hypotheses and notation of theorem 1.3.5, let $A(x) = (d_2f)(x, y(x))$ and $B(x) = (d_1f)(x, y(x))$. Then, if U is a small enough neighbourhood of a, $A(x)$ is an isomorphism for $x \in U$, $y \in C^1(U, n_2)$ and

1.3.8
$$(dy)(x) = -A(x)^{-1} \circ B(x).$$

PROOF. Since y is continuous, $x \mapsto A(x)$ is a continuous map of U_1 into the space of linear maps of R^{n_2} into itself ($n_2 \times n_2$ matrices). Further $A(a)$ is an isomorphism. Hence so is $A(x)$ for all $x \in U$, if U is a small enough neighbourhood of a. We suppose U convex. Let x, $x+\xi \in U$ and $\eta = y(x+\xi)-y(x)$. Then $f(x+\xi, y(x)+\eta) = 0$, so that, by Taylor's formula,

$$0 = f(x, y(x))+B(x)\xi+A(x)\eta + o(|\xi|+|\eta|) \text{ as } |\xi| \to 0;$$

moreover $\eta \to 0$ as $\xi \to 0$ and $f(x, y(x)) = 0$. This gives

$$A(x)\eta = -B(x)\xi+o(|\xi|+|\eta|).$$

If K is a compact subset of U, then $A(x)^{-1}$ is continuous in U, hence bounded on K (in the obvious sense), and we obtain

$$\eta = -A(x)^{-1} \circ B(x) \cdot \xi+o(|\xi|+|\eta|).$$

Hence, if $|\xi|$ is small enough, there is a constant $C > 0$ such that

$$|\eta| \leqq C|\xi| + \tfrac{1}{2}|\eta|,$$

so that $|\eta| \leqq 2C|\xi|$. Consequently

$$y(x+\xi)-y(x) = -A(x)^{-1} \circ B(x) \cdot \xi + o(|\xi|) \text{ as } |\xi| \to 0,$$

which means simply that $y \in C^1(U, n_2)$ and that (1.3.8) holds.

1.3.9 Corollary. With the hypotheses and notation of theorem 1.3.5, let U be a neighbourhood of a such that $(df)(x, y(x))$ is an isomorphism for $x \in U$. If

$$f \in C^k(\Omega_1 \times \Omega_2, n_2), \qquad k \geqq 1,$$

we have

$$y \in C^k(\Omega_2, n_2).$$

Proof. For $k = 1$, this is lemma 1.3.1. For $k > 1$, we proceed by induction. If

$$f \in C^k(\Omega_1 \times \Omega_2, n_2)$$

and

$$y \in C^r(U, n_2), \qquad r < k,$$

then, by definition $x \leftrightarrow A(x)$, $B(x)$ are C^r mappings of U into the appropriate finite-dimensional vector space. Then (1.3.8) implies that $y \in C^{r+1}(U, n_2)$.

1.3.10 Remark. With the notation as in corollary 1.3.9, if f is real analytic, then so is y. This is an immediate consequence of lemma 1.1.5 and remark 1.3.6.

1.3.11 Remark. Under the hypotheses of theorem 1.3.2, and with the same notation, if $f \in C^k(\Omega, n)$ (or is real analytic), so is $(f|U)^{-1}$. This follows at once from corollary 1.3.9 and remark 1.3.10 applied to the map $g: R^n \times \Omega \to R^n$ defined by

$$g(x, y) = x - f(y).$$

1.3.12 REMARK. Theorem 1.3.2 with the remark 1.3.11 is known as the *inverse function theorem.* The statements 1.3.5, 1.3.9 and 1.3.10 consitute the *implicit function theorem.*

1.3.13 DEFINITION. A cube in R^n is a set of the form $\{x \mid |x_j - a_j| < r_j\}$. A polycylinder in C^n is a set of the form $\{z \mid |z_j - a_j| < r_j\}$.

1.3.14 THE RANK THEOREM. Let Ω be an open set in R^n and $f \in C^k(\Omega, m)$, $k \geqq 1$, i.e. $f: \Omega \to R^m$ is a C^k map. Suppose that rank $(\mathrm{d}f)(x)$ is an integer r independent of $x \in \Omega$. Then there exist open neighbourhoods U of a and V of $b = f(a)$, cubes Q, Q' in R^n, R^m respectively and C^k diffeomorphisms $u: Q \to U$ and $u': V \to Q'$, such that if $\varphi = u' \circ f \circ u$, then φ has the form

$$\varphi(x_1, x_2, \ldots, x_n) = (x_1, x_2, \ldots, x_r, 0, \ldots, 0).$$

Further, if f is analytic, u, u' may be chosen to be, together with their inverses, also analytic.

PROOF. By affine automorphisms of R^n, R^m, we may suppose that $a = 0$, $b = 0$ and that $(\mathrm{d}f)(0)$ is the linear map

$$(v_1, \ldots, v_n) \mapsto (v_1, \ldots, v_r, 0, \ldots, 0).$$

Consider the map $w: \Omega \to R^n$ defined by

$$w(x) = (f_1(x), \ldots, f_r(x), x_{r+1}, \ldots, x_n),$$

where, of course,

$$f(x) = (f_1(x), \ldots, f_r(x), \ldots, f_m(x)).$$

Then $(\mathrm{d}w)(0)$ is the identity, hence, by the inverse function theorem (see remark 1.3.12), there exists a neighbourhood U of 0 and a cube Q such that $w|U \to Q$ is a C^k diffeomorphism. Let $u = (w|U)^{-1}$. Clearly

$$f \circ u(y) = (y_1, \ldots, y_r, \varphi_{r+1}(y), \ldots, \varphi_m(y)),$$

where the φ_j are in $C^k(Q)$. Now, if $\psi = f \circ u$, we have

$$\text{rank } (\mathrm{d}\psi)(y) = r$$

for $y \in Q$, hence

$$\frac{\partial \varphi_j}{\partial y_k} = 0 \qquad \text{if} \quad j, k > r.$$

Thus the φ_j are independent of $y_{r+1}, \ldots, y_n$. Let now $Q = Q^r \times Q^{n-r}$, where Q^r, Q^{n-r} are cubes in R^r, R^{n-r} respectively. Let

$$v: Q^r \times R^{m-r} \to Q^r \times R^{m-r}$$

be the map defined by

$$v(y_1, \ldots, y_r, \ldots, y_m) =$$
$$= (y_1, \ldots, y_r, y_{r+1} - \varphi_{r+1}(y_1, \ldots, y_r), \ldots, y_m - \varphi_m(y_1, \ldots, y_r)).$$

Trivially v is a C^k diffeomorphism. Let Q' be a cube in R^m such that

$$v \circ \psi(Q) \subset Q' \subset Q^r \times R^{m-r}$$

and let $V = v^{-1}(Q')$. If we set $u' = v|Q'$, we have

$$u' \circ f \circ u(x_1, \ldots, x_n) = u' \circ \psi(x_1, \ldots, x_n) = (x_1, \ldots, x_r, 0, \ldots, 0).$$

Note that we may replace C^k maps by real analytic ones if f is analytic.

1.3.15 Remark. It is clear that the rank theorem has an analogue for holomorphic mappings. We do not formulate this explicitly since statement and proof are practically identical with theorem 1.3.14.

§ 1.4 Sard's theorem and functional dependence

1.4.1 Lemma. Let Ω be an open set in R^n and $f\colon \Omega \to R^n$ a map which satisfies a Lipschitz condition on compact subsets of Ω, i.e., for any compact $K \subset \Omega$, there is an $M > 0$ such that $|f(x) - f(y)| \leqq M|x-y|$ for $x, y \in K$. Then, for any set $S \subset \Omega$ of measure zero, $f(S)$ has measure zero.

The proof follows at once from the definition of sets of measure zero.

1.4.2 Lemma. If Ω is open in R^n and $f \in C^1(\Omega, n)$, then f carries sets of measure zero into sets of measure zero.

1.4.3 Lemma. If m, n are integers, $m > n$ and if Ω is open in R^n and $f\colon \Omega \to R^m$ is a C^1 map, then $f(\Omega)$ has measure zero in R^m.

PROOF. If we define $g: \Omega \times R^{m-n} \to R^m$ by

$$g(x_1, \ldots, x_m) = f(x_1, \ldots, x_n),$$

then $g \circ f(\Omega) = g(\Omega \times 0)$ has measure 0 in R^m by lemma 1.4.2.

1.4.4 DEFINITION. Let Ω be an open set in R^n and $f: \Omega \to R^m$ be a C^1 map. A point $a \in \Omega$ is called a critical point of f if rank $(\mathrm{d}f)(a) < m$.

1.4.5 REMARK. (a) If $m > n$, every point of Ω is a critical point. (b) The set of critical points of f is closed in Ω.

The main object of this section is to prove the following theorem.

1.4.6 THEOREM OF SARD. If Ω is an open set in R^n, $f: \Omega \to R^m$ is a C^∞ map and if A is the set of critical points of f, then $f(A)$ has measure zero in R^m.

As a matter of fact, the theorem is true if we suppose merely that $f \in C^r(\Omega, m)$, where $r = \max(n-m+1, 1)$. The proof of this stronger version requires, however, somewhat more delicate analysis: see SARD [1942] and MORSE [1939]; also MALGRANGE [1966]. WHITNEY [1936] has shown that the differentiability requirement max $(n-m+1, 1)$ is the best possible; he has given an example of a C^{n-m} map f $(n > m)$ from R^n to R^m for which the image of the set of critical points contains a non-empty open set.

We shall begin by proving the Sard theorem when $m = n$, but under the weaker differentiability assumption, before taking up the proof of theorem 1.4.6.

1.4.7 PROPOSITION. Let Ω be an open set in R^n and $f: \Omega \to R^n$ a C^1 map If A is the set of critical points of f, then $f(A)$ has measure 0 in R^n.

PROOF. Let $a \in A$. By assumption $\operatorname{rank}(\mathrm{d}f)(a) < n$; hence $f(a) + (\mathrm{d}f)(a)(R^n)$ is an affine subspace V_a of R^n of dimension $< n$. Let $u_1, \ldots, u_n$ be an orthonormal basis for R^n with centre $f(a)$, such that V_a lies in the space spanned by $u_1, \ldots, u_{n-1}$. Let Q be a closed cube

$\subset \Omega$; we have only to show that $f(Q \cap A)$ has measure 0. Now, if $x, y \in Q$, we have

$$f(x)-f(y) = (\mathrm{d}f)(y)(x-y)+r(x, y),$$

where

$$r(x, y) = \mathrm{o}(||x-y||),$$

uniformly on $Q \times Q$, as $||x-y|| \to 0$; hence there is a function $\lambda: R^+ \to R^+$ with $\lambda(t) \to 0$ as $t \to 0$ such that

$$||r(x, y)|| \leqq \lambda(||x-y||)||x-y||.$$

If $\varepsilon > 0$ is sufficiently small, and if x lies in a cube Q_ε of side ε containing $a \in A, f(x)$ lies in the region between the hyperplanes

$$u_n = 2\lambda(\varepsilon)\,\varepsilon \quad \text{and} \quad u_n = -2\lambda(\varepsilon)\varepsilon.$$

Moreover, by Taylor's formula, $f(x)$ lies in a cube of side $M\varepsilon$ and centre $f(a)$ (with sides parallel to the coordinate axes with respect to the u_i), M being a constant independent of a, x and the choice of coordinates $\{u_i\}$. The volume of the intersection of the cube of side $M\varepsilon$ and the region between the hyperplanes $u_n = \pm 2\lambda(\varepsilon)\varepsilon$ is $\leqq 4M^n\varepsilon^n\lambda(\varepsilon)$. Since an orthonormal change of coordinates leaves the measure in R^n invariant, we see that $f(Q_\varepsilon)$ has measure $\leqq 4M^n\varepsilon^n\lambda(\varepsilon)$. Let l be the length of a side of Q. Divide Q into $(l/\varepsilon)^n$ cubes of side ε, $i = 1, \ldots, (l/\varepsilon)^n$. We have seen that if $Q_i \cap A \neq \emptyset$, then

$$\text{measure}\, f(Q_i) \leqq 4M^n\varepsilon^n\lambda(\varepsilon).$$

Hence

$$\text{measure}\, f(A \cap Q) \leqq \sum_{i,\, A\cap Q_i \neq \phi} \text{measure}\, f(A \cap Q_i) \leqq l^n 4M^n\lambda(\varepsilon).$$

Since $\lambda(\varepsilon) \to 0$ as $\varepsilon \to 0$, measure $f(A \cap Q) = 0$.

For the proof of theorem 1.4.6, we need a preliminary proposition.

1.4.8 PROPOSITION. If $f: \Omega \to R^1$ is a C^∞ function and A is the set of critical points of f, then $f(A)$ has measure 0.

PROOF. Define

$$A_k = \{a \in \Omega \mid D^\alpha f(a) = 0 \text{ for all } \alpha \text{ with } 0 < |\alpha| \leqq k\}.$$

Clearly $A_{k+1} \subset A_k$ and we have

1.4.9 $\quad A = A_1 = (A_1 - A_2) \cup (A_2 - A_3) \cup \ldots \cup (A_{n-1} - A_n) \cup A_n.$

Now, if $a \in A_n$ and Q is any closed cube in Ω with $a \in \Omega$, we have

$$|f(x) - f(a)| \leqq M|x-a|^{n+1},$$

so that the image of a cube of side ε about a has measure $\leqq M\varepsilon^{n+1}$ in R^1($M > 0$ a fixed constant). Hence, by splitting Q into $(l/\varepsilon)^n$ cubes of side ε, we see, as in the proof of proposition 1.4.7, that $f(A_n)$ has a measure $< l^n M\varepsilon$, and since ε is arbitrary, $f(A_n)$ has measure 0. If $n = 1$, then $A = A_1 = A_n$, so that proposition 1.4.8 is proved in this case. We now suppose, by induction, that if Ω' is an open set in R^{n-1} and g is a C^∞ map of Ω' into R^1, and if A_g is the set of critical points of g, then $g(A_g)$ has measure zero.

Returning to the decomposition (1.4.9) relative to our map f: $\Omega \to R^1$, we have only to show that $f(A_k - A_{k+1})$ has measure 0 for $1 \leqq k < n$. For this, if we set $B_k = A_k - A_{k+1}$, it suffices to show that any $a \in B_k$ has a neighbourhood U in R^n such that $f(U \cap B_k)$ has measure 0.

Since $a \notin A_{k+1}$, there is a multi-index

$$\alpha = (\alpha_1, \ldots, \alpha_n), \qquad |\alpha| = k+1,$$

such that

$$D^\alpha f(a) \neq 0.$$

If $\alpha_j \neq 0$, let

$$\beta = \alpha - (0, \ldots, 1, \ldots, 0),$$

with 1 in the jth place and 0 elsewhere, and let $h = D^\beta f$. Then $(\mathrm{d}h)(a)$ has maximal rank 1 at a, so that, by the rank theorem 1.3.14, there is a neighbourhood U of a, a cube Q in R^n and a C^∞ diffeomorphism u: $U \to Q$ such that

$$u(\{x \mid h(x) = 0\}) = \{(x_1, \ldots, x_n) \in Q \mid x_1 = 0\} = H \text{ say.}$$

By hypothesis, $u(B_k) \subset H$, and we set

$$\Omega' = \{(x_2, \ldots, x_n) \in R^{n-1} | (0, x_2, \ldots, x_n) \in H\}.$$

Let g be the C^∞ function on Ω' defined by

$$g(x_2, \ldots, x_n) = F(0, x_2, \ldots, x_n),$$

where $F = f \circ u^{-1}$. If $S = u(B_k \cap U)$, we clearly have $F(S) \subset g(A_g)$, where A_g is the set of critical points of g. By inductive hypothesis $g(A_g)$, hence also $F(S) = f(U \cap B_k)$, has measure zero.

1.4.10 COROLLARY. If $f: \Omega \to R^m$ is a C^∞ map and $B = \{x | (\mathrm{d}f)(x) = 0\}$, then $f(B)$ has measure 0 in R^m.

PROOF. If $f = (f_1, \ldots, f_m)$, then $B \subset B_1 = \{x | (\mathrm{d}f_1)(x) = 0\}$; hence

$$f(B) \subset f_1(B_1) \times R^{m-1}.$$

By proposition 1.4.8, $f_1(B_1)$ has measure zero in R, so that $f_1(B_1) \times R^{m-1}$ has measure zero in R^m.

We require finally one more result which is an immediate consequence of the theorem of Fubini on the representation of double integrals as iterated integrals.

1.4.11 LEMMA. Let S be a measurable set in $R^p = R^r \times R^{p-r}$ $(0 < r < p)$. We denote a point in R^p by (x, y), $x \in R^r$, $y \in R^{p-r}$. For $c \in R^r$, let

$$S_c = \{y \in R^{p-r} | (c, y) \in S\}.$$

Then S has measure zero in R^p if and only if S_c has measure zero in R^{p-r} for almost all $c \in R^p$.

One has only to apply Fubini's theorem to the characteristic function of S [which is, by definition, 1 on S and 0 outside].

PROOF OF THEOREM 1.4.6. Let

$$E_k = \{x \in \Omega | \operatorname{rank} (\mathrm{d}f)(x) = k\},$$

$f: \Omega \to R^m$, being the given C^∞ map. If $m > n$, the theorem is an immediate consequence of lemma 1.4.3. We may therefore suppose that $n \geqq m$. We then have

$$A = \bigcup_{0 \leqq k < m} E_k .$$

We have to show that any $a \in E_k$ has a neighbourhood U in R^n such that $f(U \cap E_k)$ has measure zero. Now, the set

$$\{x \in \Omega | \operatorname{rank}(\mathrm{d}f)(x) \leqq k\} = \bigcup_{0 \leqq r \leqq k} E_r$$

is closed for each k; hence E_k is locally closed, i.e., for any $a \in E_k$ and all small enough neighbourhoods U of a in R^n, $U \cap E_k$ is closed in U, hence a countable union of compact sets. Since the image of a compact set under f is compact, hence measurable, $S_k = f(U \cap E_k)$ is measurable in R^m.

If $k = 0$, S_0 has measure zero by corollary 1.4.10. Let $0 < k < m$ and $a \in E_k$. If $f = (f_1, \ldots, f_m)$, we may suppose, by a permutation of the f_j, that

$$\operatorname{rank}(\mathrm{d}u)(a) = k,$$

where $u = (f_1, \ldots, f_k)$. There exist C^∞ functions $u_{k+1}, \ldots, u_n$ on Ω (in fact linear functions on R^n) so that

$$\operatorname{rank}(\mathrm{d}w)(a) = n,$$

where

$$w = (f_1, \ldots, f_k, u_{k+1}, \ldots, u_n).$$

By the inverse function theorem 1.3.11, there are arbitrarily small neighbourhoods U of a and V of $w(a)$ such that $w: U \to V$ is a C^∞ diffeomorphism. The map

$$F = f \circ w^{-1}: V \to R^m$$

has the form

$$F(u_1, \ldots, u_n) = (u_1, \ldots, u_k, F_{k+1}(u), \ldots, F_m(u));$$

if

$$E_k' = \{u \in V \mid (\mathrm{d}F)(u) \text{ has rank } k\},$$

then

$$S_k = f(U \cap E_k) = F(E_k').$$

For $c \in R^k$, we define the map

$$F_c\colon V_c \to R^{m-k}$$

by

$$F_c(y) = (F_{k+1}(c, y), \ldots, F_m(c, y)).$$

Here $V_c = \{y \in R^{n-k} \mid (c, y) \in V\}$. It is clear that

$$(c, y) \in E_k' \Leftrightarrow (dF_c)(y) = 0.$$

Hence, by corollary 1.4.10, if

$$E_{k,c}' = \{y \in V_c \mid (c, y) \in E_k'\},$$

we have $F_c(E_{k,c}')$ is of measure zero in R^{m-k}. Further

$$S_{k,c} = \{y \in R^{m-k} \mid (c, y) \in S_k = F(E_k')\} = F_c(E_{k,c}')\}.$$

Hence, since S_k is measurable, S_k has measure 0 in R^m by lemma 1.4.11. As we have already remarked, this completes the proof of theorem 1.4.6.

The theorem, in its general form, is due to Marston Morse, A. P. Morse and Sard.

We give now an application of Sard's theorem.

1.4.12 Application. Let $f_1, \ldots, f_m \in C^\infty(\Omega)$. The $\{f_j\}$ are said to be functionally dependent on a subset S of Ω if there exists an open set

$$\Omega' \supset f(S), \{f = (f_1, \ldots, f_m)\colon \Omega \to R^m\}$$

and a C^∞ function g on Ω' such that $g^{-1}(0)$ is nowhere dense in Ω' and $g(f(x)) = 0$ for $x \in S$. If g can be chosen real analytic, the $\{f_j\}$ are called analytically dependent. A corresponding definition applies to holomorphic functions.

1.4.13 Lemma. If X is any closed set in R^n, there is a $\varphi \in C^\infty(R^n)$ such that $X = \{x \in R^n \mid \varphi(x) = 0\}$.

PROOF. There exist open sets $U_p, p \geqq 1$ in R^n such that $X = \bigcap_{p \geqq 1} U_p$. Let $\{K_m\}$ be a sequence of compact subsets of R^n such that

$$\bigcup_{m=1}^{\infty} K_m = R^n, \qquad K_m \subset \mathring{K}_{m+1}.$$

By corollary 1.2.6, there is a $\varphi_p \in C^\infty(R^n)$ such that $0 \leqq \varphi_p \leqq 1$ and $\varphi_p = 0$ on X and $\varphi_p = 1$ on $R^n - U_p$.

Let

$$c_p = \|\varphi_p\|_p^{K_p} = \sum_{|\alpha| \leqq p} \frac{1}{\alpha!} \sup_{x \in K_p} |D^\alpha \varphi_p(x)|.$$

Chose $\varepsilon_p > 0$ such that

$$\sum_1^\infty \varepsilon_p c_p < \infty.$$

Let

$$\psi_m = \sum_{p=1}^{m} \varepsilon_p \varphi_p.$$

Then for any compact set $K \subset R^n$, we have $K \subset K_r$ for some r, so that if $p > 0$ is given and $m' \geqq m > r, p$, then

$$\|\psi_m - \psi_{m'}\|_p^K \leqq \sum_{q>m} \varepsilon_q \|\varphi_q\|_p^K \leqq \sum_{q>m} \varepsilon_q \|\varphi_q\|_q^{K_q} \to 0 \quad \text{as} \quad m \to \infty.$$

Hence $\{\psi_m\}$ is a Cauchy sequence in $C^\infty(R^n)$, and its limit φ clearly has the required properties.

1.4.14 THEOREM. If $f: \Omega \to R^m$ is a C^∞ map, $f = (f_1, \ldots, f_m)$, then $\{f_j\}$ are functionally dependent on every compact subset of Ω if and only if rank $(\mathrm{d}f)(x) < m$ for all $x \in \Omega$.

PROOF. Suppose that for some $a \in \Omega$, rank $(\mathrm{d}f)(a) = m$. Then rank $(\mathrm{d}f)(x) = m$ for all x sufficiently near a, so that, by the rank theorem, there is a relatively compact neighbourhood U of a such that $f(U)$ is open in R^m, so that $f(\bar{U})$ is not nowhere dense. Obviously the $\{f_j\}$ are not functionally dependent on $\bar{U}$.

Conversely if rank $(\mathrm{d}f)(x) < m$ for all $x \in \Omega$, by Sard's theorem 1.4.6, $f(\Omega)$ has measure 0 in R^m. If $K \subset \Omega$ is compact, $f(K)$ is a compact set of measure 0, hence is nowhere dense. By lemma 1.4.13, there is a $g \in C^\infty(R^m)$ with $g^{-1}(0) = f(K)$. Clearly $g \circ f(x) = 0$ for $x \in K$.

Only a somewhat weaker statement is true of analytic dependence.

1.4.15 THEOREM. Let $f\colon \Omega \to R^m$ be an analytic map, $f = (f_1, \ldots, f_m)$. Then rank $(\mathrm{d}f)(x) < m$ for any $x \in \Omega$ if and only if the following holds: there exists a nowhere dense closed set $S \subset \Omega$ with the property that any $a \in \Omega - S$ has a neighbourhood $U \subset \Omega$ such that the $f_j|U$ are analytically dependent.

Note that, by the principle of analytic continuation, if Ω is connected, then rank $(\mathrm{d}f)(x) < m$ *for all* $x \in \Omega$ if and only if this is the case for all x in a non-empty open subset of Ω.

PROOF. We may suppose that Ω is connected. If a set S with the properties stated above exists, then clearly rank $(\mathrm{d}f)(x) < m$ for $x \in \Omega - S$ (theorem 1.4.14) and so on Ω, since the set

$$\{x \mid \operatorname{rank} (\mathrm{d}f)(x) = m\}$$

is open.

Conversely, let

$$p = \max_x \operatorname{rank} (\mathrm{d}f)(x) < m;$$

choose $b \in \Omega$ with

$$\operatorname{rank} (\mathrm{d}f)(b) = p.$$

This implies that there exist indices $j_1, \ldots, j_p$, $1 \leqq j_r \leqq m$ and $k_1, \ldots, k_p$, $1 \leqq k_r \leqq n$ such that $h(b) \neq 0$, where

$$h(x) = \det \left(\frac{\partial f_{j_r}}{\partial x_{k_s}}(x) \right).$$

Let $S = \{x \in \Omega \mid h(x) = 0\}$. Since h is analytic in Ω and $\not\equiv 0$, S can contain no open set, so is nowhere dense.

Obviously,

$$\operatorname{rank} (\mathrm{d}f)(x) = p, \qquad x \in \Omega - S.$$

By the rank theorem 1.3.15, there exist neighbourhoods U of a, V of $f(a)$, cubes Q, Q' in R^n, R^m respectively and analytic isomorphisms $u: Q \to U$, $u': V \to Q'$ such that $u' \circ f \circ u$ is the map $(x_1, \ldots, x_n) \mapsto (x_1, \ldots, x_p, 0, \ldots, 0)$. If $u' = (u'_1, \ldots, u'_m)$ and we set $g = u'_m$, we have

$$g \circ f = 0 \quad \text{on} \quad U.$$

1.4.16 EXAMPLE. Let $\varphi(z)$ be an entire function of the complex variable z, not a polynomial and real on the real axis (e.g., $\varphi(z) = \exp(z)$). Consider the map $f: R^2 \to R^3$ given by

$$f(x_1, x_2) = (x_1, x_1 x_2, x_1 \varphi(x_2)).$$

It can be shown that there does not exist an analytic function $g \not\equiv 0$ in a neighbourhood of $0 \in R^3$ such that $g \circ f = 0$ in a neighbourhood of $0 \in R^2$. This shows that the presence of the set S in theorem 1.4.15 is necessary.

1.4.17 REMARK. Theorem 1.4.15 and the example 1.4.16 apply also to holomorphic functions with the obvious changes.

§ 1.5 Borel's theorem on Taylor series

Let Ω be an open set in R^n such that $0 \in \Omega$ and let $f \in C^\infty(\Omega)$. We denote by $T(f)$ the formal power series

$$T(f) = \sum_{\alpha} \frac{1}{\alpha!} (D^\alpha f)(0) x^\alpha;$$

if $m > 0$ is an integer, we set

$$T^m(f) = \sum_{|\alpha| \leqq m} \frac{1}{\alpha!} (D^\alpha f)(0) x^\alpha;$$

$T^m(f)$ is, of course, a polynomial.

1.5.1 DEFINITION. Let X be a closed subset of Ω. We say that $f \in C^k(\Omega)$ is m-flat on X, $m \leqq k$, if $D^\alpha f(x) = 0$ for all $x \in X$ and all α with $|\alpha| \leqq m$. If $f \in C^\infty(\Omega)$ and $D^\alpha f(x) = 0$ for $x \in X$ and all α, we say that f is flat on X.

1.5.2 LEMMA. Let $f \in C^\infty(R^n)$ be m-flat at 0. Then, given $\varepsilon > 0$, there exists $g \in C^\infty(R^n)$ which vanishes in a neighbourhood of 0 and such that

$$\|g-f\|_m^{R^n} < \varepsilon.$$

PROOF. By corollary 1.2.6, there is an $\eta \in C^\infty(R^n)$ such that $\eta(x) \geqq 0$ for all x and $\eta(x) = 0$ if $|x| \leqq \frac{1}{2}$, $\eta(x) = 1$ if $|x| \geqq 1$. For $\delta > 0$, define

$$g_\delta(x) = \eta\left(\frac{x}{\delta}\right) f(x).$$

Clearly $g_\delta \in C^\infty(R^n)$ and vanishes near 0; it is therefore sufficient to prove that

$$\sup_{x \in R^n} |(D^\alpha g_\delta)(x) - (D^\alpha f)(x)| \to 0 \quad \text{as} \quad \delta \to 0 \qquad \text{if} \quad |\alpha| \leqq m.$$

Now,

$$g_\delta(x) = f(x) \qquad \text{if} \quad |x| \geqq \delta;$$

hence

$$\sup_{x \in R^n} |D^\alpha g_\delta(x) - D^\alpha f(x)| = \sup_{|x| \leqq \delta} |D^\alpha g_\delta(x) - D^\alpha f(x)|.$$

Since f is m-flat at 0, $D^\alpha f(0) = 0$, so that

$$\sup_{|x| \leqq \delta} |D^\alpha f(x)| \to 0 \quad \text{as} \quad \delta \to 0 \quad \text{if} \quad |\alpha| \leqq m.$$

Now we have

$$D^\alpha g_\delta(x) = \sum_{\mu+\nu=\alpha} \binom{\alpha}{\nu} \delta^{-|\nu|} (D^\nu \eta)\left(\frac{x}{\delta}\right) (D^\mu f)(x).$$

Since $\eta(x) = 1$ if $|x| \geqq 1$,

$$\sup_{x \in R^n} |D^\nu \eta(x)| = M_\nu < \infty.$$

Hence

$$|D^\alpha g_\delta(x)| \leqq M \sum_{\mu+\nu=\alpha} \delta^{-|\nu|} |D^\mu f(x)|, \qquad M = \max_\nu \binom{\alpha}{\nu} M_\nu.$$

Now, $D^\mu f$ is $(m-|\mu|)$-flat at 0, so that

$$\sup_{|x|\leq\delta} |D^\mu f(x)| = o(\delta^{m-|\mu|}) \quad \text{as} \quad \delta \to 0;$$

hence

$$\sup_{|x|\leq\delta} |D^\alpha g_\delta(x)| = o(\sum_{\mu+\nu=\alpha} \delta^{m-|\mu|-|\nu|}) = o(1) \quad \text{if} \quad |\alpha| \leq m.$$

This proves the lemma.

1.5.3 REMARK. Actually, it is sufficient to suppose that $f \in C^m(R^n)$ and is m-flat at 0. The function g of lemma 1.5.2 is, in particular, flat at 0.

1.5.4 THEOREM OF BOREL. Given, for each n-tuple $\alpha = (\alpha_1, \ldots, \alpha_n)$ of non-negative integers, a real constant c_α, there exists an $f \in C^\infty(R^n)$ such that

$$\frac{1}{\alpha!} D^\alpha f(0) = c_\alpha.$$

In other words, the mapping from $C^\infty(R^n)$ to the ring of formal power-series in n-variables given by $f \mapsto T(f)$ is surjective.

PROOF. Let

$$T_m(x) = \sum_{|\alpha|\leq m} c_\alpha x^\alpha.$$

Clearly $T_{m+1} - T_m$ is m-flat at 0, so that, by lemma 1.5.2, there exists a $g_m \in C^\infty(R^n)$ vanishing in a neighbourhood of 0 such that

$$\|T_{m+1} - T_m - g_m\|_m^{R^n} < 2^{-m}.$$

Clearly, the function

$$f = T_0 + \sum_{m=0}^{\infty} (T_{m+1} - T_m - g_m) \in C^\infty(R^n).$$

Further, for any $k > 0$, the sum $\sum_{m\geq k}(T_{m+1} - T_m - g_m)$ is k-flat at 0. Hence

$$T^k(f) = T^k(T_0 + \sum_{m=0}^{k-1} (T_{m+1} - T_m - g_m)) = T_k, \qquad \text{q.e.d.}$$

This theorem of Borel is a very special case of important theorems of WHITNEY [1934] on differentiable functions on closed sets. We shall next state, without proof, one of Whitney's main theorems in this direction. An elegant proof is given in the book of MALGRANGE [1966], based on simplifications introduced by GLAESER [1958].

1.5.5 EXTENSION THEOREM OF WHITNEY: PART 1. Let k be an integer > 0, Ω an open set in R^n and X a closed subset of Ω. Suppose that for each n-tuple $\alpha = (\alpha_1, \ldots, \alpha_n)$ of non-negative integers, $|\alpha| \leqq k$, there is given a continuous function f_α on X. Then the necessary and sufficient condition that there exist $f \in C^k(\Omega)$ for which $D^\alpha f|X = f_\alpha$, $|\alpha| \leqq k$, is that for any α, $|\alpha| \leqq k$, we have

$$f_\alpha(x) = \sum_{|\beta| \leqq k-|\alpha|} \frac{1}{\beta!} f_{\alpha+\beta}(y)(x-y)^\beta + \mathrm{o}(|x-y|^{k-|\alpha|}),$$

uniformly on compact subsets of X as $|x-y| \to 0$.

1.5.6 EXTENSION THEOREM OF WHITNEY: PART 2. Given a continuous function f_α on X for each α, there exists an $f \in C^\infty(\Omega)$ with $D^\alpha f|X = f_\alpha$ for all α if and only if for any integer $m > 0$ and any compact set $K \subset X$ we have

$$f_\alpha(x) = \sum_{|\beta| \leqq m} \frac{1}{\beta!} f_{\alpha+\beta}(y)(x-y)^\beta + \mathrm{o}(|x-y|^m),$$

uniformly as $|x-y| \to 0$, $x, y \in K$.

Borel's theorem is the special case of theorem 1.5.6 when $X = \{0\}$ reduces to the single point 0.

§ 1.6 Whitney's approximation theorem

We begin with a lemma.

1.6.1 LEMMA. Let $f \in C_0^k(R^n)$, $0 \leqq k < \infty$. For $\lambda > 0$, set

$$\begin{aligned} I_\lambda(f)(x) &\equiv g_\lambda(x) \\ &= c\lambda^{\frac{1}{2}n} \int_{R^n} f(y) \exp\{-\lambda[(x_1-y_1)^2 + \ldots + (x_n-y_n)^2]\}\mathrm{d}y \\ &\equiv c\lambda^{\frac{1}{2}n} \int_{R^n} f(y) \exp\{-\lambda\|x-y\|^2\}\mathrm{d}y, \end{aligned}$$

where $c = \pi^{-\frac{1}{2}n}$, so that

$$c\int_{R^n} \exp(-||x||^2)\mathrm{d}x = 1.$$

Then, we have

$$||g_\lambda - f||_k^{R^n} \to 0 \qquad \text{as} \quad \lambda \to \infty.$$

Note that $g_\lambda \in C^\infty(R^n)$.

PROOF. We have

$$g_\lambda(x) = c\lambda^{\frac{1}{2}n}\int_{R^n} f(x-y)\exp(-\lambda||y||^2)\mathrm{d}y.$$

so that, for $|\alpha| \leqq k$, we have

$$D^\alpha g_\lambda(x) = c\lambda^{\frac{1}{2}n}\int_{R^n} (D^\alpha f)(x-y)\exp(-\lambda||y||^2)\mathrm{d}y$$

$$= c\lambda^{\frac{1}{2}n}\int_{R^n} (D^\alpha f)(y)\exp(-\lambda||x-y||^2)\mathrm{d}y.$$

This gives

$$D^\alpha g_\lambda(x) - D^\alpha f(x) = c\lambda^{\frac{1}{2}n}\int_{R^n} \{D^\alpha f(y) - D^\alpha f(x)\}\exp(-\lambda||x-y||^2)\mathrm{d}y.$$

Given $\varepsilon > 0$, there exists a $\delta > 0$ such that

$$|D^\alpha f(y) - D^\alpha f(x)| < \varepsilon/2 \qquad \text{for} \quad ||x-y|| \leqq \delta \quad \text{and} \quad |\alpha| \leqq k,$$

since $f \in C_0^k(R^n)$.

Moreover, there is an $M > 0$ such that

$$|D^\alpha f(y)| < M \text{ for all } y, \qquad |\alpha| \leqq k.$$

Hence

$$|D^\alpha g_\lambda(x) - D^\alpha f(x)| =$$

$$c\lambda^{\frac{1}{2}n}\left|\left(\int_{||x-y||<\delta} + \int_{||x-y||\geqq\delta}\right)\{D^\alpha f(y) - D^\alpha f(x)\}\exp(-\lambda||x-y||^2)\mathrm{d}y\right|$$

$$\leqq \tfrac{1}{2}\varepsilon c\lambda^{\frac{1}{2}n}\int_{R^n} \exp(-\lambda||x-y||^2)\mathrm{d}y$$

$$+ 2Mc\lambda^{\frac{1}{2}n}\int_{||x-y||\geqq\delta} \exp(-\lambda||x-y||^2)\mathrm{d}y.$$

Now,

$$c\lambda^{\frac{1}{2}n}\int_{R^n}\exp(-\lambda||x-y||^2)dy = 1$$

and

$$c\lambda^{\frac{1}{2}n}\int_{||x-y||\geqq\delta}\exp(-\lambda||x-y||^2)dy$$
$$\leqq e^{-\frac{1}{2}\lambda\delta^2}c\lambda^{\frac{1}{2}n}\int_{R^n}\exp(-\tfrac{1}{2}\lambda||x-y||^2)dy = 2^{\frac{1}{2}n}\exp(-\tfrac{1}{2}\lambda\delta^2).$$

This gives

$$|D^\alpha g_\lambda(x)-D^\alpha f(x)| \leqq \tfrac{1}{2}\varepsilon+M2^{1+\frac{1}{2}n}\exp(-\tfrac{1}{2}\lambda\delta^2).$$

Since, for fixed $\delta > 0$, we can choose λ so large that the term on the right-hand side is $< \varepsilon$, we see that

$$\sup_{x\in R^n}|D^\alpha g_\lambda(x)-D^\alpha f(x)| \to 0 \qquad \text{as} \quad \lambda\to\infty,$$

which proves the lemma.

1.6.2 THE WEIERSTRASS APPROXIMATION THEOREM. Let Ω be open in R^n. Then given any $f\in C^k(\Omega)$, $0 \leqq k < \infty$, $\varepsilon > 0$ and a compact set $K \subset \Omega$, there is a polynomial $P(x)$ in $x_1, \ldots, x_n$ such that

$$||f-P||_k^K < \varepsilon.$$

PROOF. Replacing f by φf where $\varphi\in C_0^\infty(\Omega)$, supp $(\varphi)\subset\Omega$ and which $= 1$ in a neighbourhood of K [such a φ exists by corollary 1.2.6], we may suppose that $f\in C_0^k(R^n)$. Then, by lemma 1.6.1, given $\varepsilon > 0$, we can choose λ so large that if

$$I_\lambda(f)(x) \equiv g_\lambda(x) = c\lambda^{\frac{1}{2}n}\int_{R^n}f(y)\exp(-\lambda||x-y||^2)dy,$$

we have

$$||g_\lambda-f||_k^K < \tfrac{1}{2}\varepsilon.$$

Now

$$\exp(-\lambda||x-y||^2) = \sum_{p=0}^{\infty}\frac{1}{p!}(-\lambda)^p||x-y||^{2p}.$$

If we set

$$Q_N(x, y) = \sum_{p=0}^{N} \frac{1}{p!}(-\lambda)^p\|x-y\|^{2p},$$

then

$$D^\alpha Q_N(x, y) \to D^\alpha \exp(-\lambda\|x-y\|^2)$$

uniformly for x, y in any compact set. If then

$$P_N(x) = c\lambda^{\frac{1}{2}n}\int f(y)Q_N(x, y)\mathrm{d}y,$$

P_N is a polynomial and $\|g_\lambda - P_N\|_k^K \to 0$ as $N \to \infty$.

1.6.3 COROLLARY. If Ω_j is an open set in R^{n_j}, $j = 1, 2$ and x_j denotes a general point in R^{n_j}, the finite linear combinations

$$\sum_\mu \varphi_\mu^{(1)}(x_1)\varphi_\mu^{(2)}(x_2),$$

where $\varphi_\mu^{(j)} \in C^\infty(R^{n_j})$, are dense in $C^k(\Omega_1 \times \Omega_2)$ for $0 \leqq k \leqq \infty$.

Since the topology on $C^k(\Omega_1 \times \Omega_2)$ involves only approximation on compact sets, multiplying the $\varphi_\mu^{(j)}$ by suitable C^∞ functions with compact support we obtain.

1.6.4 COROLLARY. With the notation of corollary 1.6.3, the finite linear combinations $\sum_\mu \varphi_\mu^{(1)}(x_1)\varphi_\mu^{(2)}(x_2)$, where $\varphi_\mu^{(j)} \in C_0^\infty(\Omega_j)$, are dense in $C^k(\Omega_1 \times \Omega_2)$ for $0 \leqq k \leqq \infty$.

We now take up an approximation theorem of WHITNEY [1934] which is of fundamental importance in the finer study of differentiable and analytic manifolds.

1.6.5 WHITNEY'S APPROXIMATION THEOREM. Let Ω be open in R^n and $f \in C^k(\Omega)$, $0 \leqq k \leqq \infty$. Let η be a continuous function on Ω with $\eta(x) > 0$ for any $x \in \Omega$. Then there exists a real analytic function g on Ω such that we have

$$|D^\alpha f(x) - D^\alpha g(x)| < \eta(x) \quad \text{for} \quad 0 \leqq |\alpha| \leqq \min\left(k, \frac{1}{\eta(x)}\right);$$

(of course, $\min(\infty, a) = a$ if $a \in R$).

We begin by reformulating the result a little.

1.6.6 REMARK. Let $\{K_p\}$ be a sequence of compact subsets of Ω such that $K_0 = \emptyset$, $K_p \subset \mathring{K}_{p+1}$ and $\bigcup K_p = \Omega$ If $\{\varepsilon_p\}$ is a sequence of strictly positive numbers, there exists a continuous function η with $\eta(x) > 0$ for any x and $\eta(x) < \varepsilon_p$ for $x \in K_{p+1} - K_p$, $p \geqq 0$.

Hence, theorem 1.6.5 can be stated as follows.

1.6.7 THEOREM. Let Ω be open in R^n and $f \in C^k(\Omega)$, $0 \leqq k \leqq \infty$. Let $\{K_p\}$ be a sequence of compact sets in Ω with $K_0 = \emptyset$, $K_p \subset \mathring{K}_{p+1}$ and $\bigcup K_p = \Omega$. Let $\{n_p\}$ be an arbitrary sequence of positive integers, and let $m_p = \min(k, n_p)$. Finally, let $\{\varepsilon_p\}$ be any sequence of numbers > 0. Then there exists a real analytic function g on Ω such that

$$\|f-g\|_{m_p}^{K_{p+1}-K_p} < \varepsilon_p,$$

for every $p \geqq 0$.

PROOF. We may suppose that $m_{p+1} \geqq m_p$ for $p \geqq 0$. Recall (1.1.10) that if $\varphi, \psi \in C^{m_p}(\Omega)$ and $S \subset \Omega$, we have

$$\|\varphi\psi\|_{m_p}^S \leqq \|\varphi\|_{m_p}^S \|\psi\|_{m_p}^S.$$

Set $L_p = K_{p+1} - K_p$ $(p \geqq 0)$. Let $\varphi_p \in C^\infty(\Omega)$ be such that $\operatorname{supp}(\varphi_p)$ is compact in Ω, $\varphi_p(x) = 0$ if x is in a neighbourhood of K_{p-1}, $\varphi_p(x) = 1$ if x is in a neighbourhood of $\bar{L}_p$. [The φ_p exist by corollary 1.2.6]. Let

$$M_p = 1 + \|\varphi_p\|_{m_p}^\Omega.$$

Let $\delta_p > 0$ be so chosen that

1.6.8 $$2\delta_{p+1} \leqq \delta_p, \qquad \sum_{q \geqq p} \delta_q M_{q+1} \leqq \tfrac{1}{4}\varepsilon_p \qquad \text{for} \quad p \geqq 0.$$

As in lemma 1.6.1, if $f \in C_0^0(R^n)$, we define $I_\lambda(f)$ by

$$I_\lambda(f)(x) = c\lambda^{\frac{1}{2}n} \int_{R^n} f(y) \exp(-\lambda\|x-y\|^2) dy,$$

$$c \int_{R^n} \exp(-\|x\|^2) dx = 1.$$

By lemma 1.6.1, there exists a $\lambda_0 > 0$ such that if $g_0 = I_{\lambda_0}(\varphi_0 f)$, we have

$$||g_0 - \varphi_0 f||_{m_0}^{K_1} < \delta_0 .$$

We now define, inductively, numbers $\lambda_0, \ldots, \lambda_p, \ldots$ and functions $g_0, g_1, \ldots, g_p, \ldots$ as follows. Suppose $g_0, \ldots, g_{p-1}, \lambda_0, \ldots, \lambda_{p-1}$ already given. By lemma 1.6.1, there is a function $l_p(\lambda_j, g_j), j \leqq p-1$, such that, if $\lambda_p > l_p$ and $g_p = I_{\lambda_p}[\varphi_p(f - g_0 - \ldots - g_{p-1})]$, then

1.6.9 $$||g_p - \varphi_p(f - g_0 - \ldots - g_{p-1})||_{m_p}^{K_{p+1}} < \delta_p .$$

Since g_p depends only on λ_p and $g_0, \ldots, g_{p-1}$, we see that l_p is a function of $\lambda_0, \ldots, \lambda_{p-1}$, only.

Since $\varphi_p = 0$ on a neighbourhood of K_{p-1}, (1.6.9) implies in particular

1.6.10 $$||g_p||_{m_p}^{K_{p-1}} < \delta_p ;$$

since $\varphi_p = 1$ on a neighbourhood of $\bar{L}_p$, we have, moreover,

1.6.11 $$||f - g_0 - \ldots - g_p||_{m_p}^{L_p} < \varepsilon_p , \qquad L_p = K_{p+1} - K_p .$$

Thus (1.6.9), with p replaced by $p+1$ gives us

$$\begin{aligned} ||g_{p+1}||_{m_p}^{L_p} &\leqq ||\varphi_{p+1}(f - \sum_0^p g_q)||_{m_p}^{L_p} + ||g_{p+1} - \varphi_{p+1}(f - \sum_0^p g_q)||_{m_p}^{L_p} \\ &\leqq ||\varphi_{p+1}||_{m_p}^{\Omega} ||f - \sum_0^p g_q||_{m_p}^{L_p} + \delta_{p+1} \\ &\leqq M_{p+1}\delta_p + \delta_{p+1} ; \end{aligned}$$

moreover, we have (1.6.10)

$$||g_{p+1}||_{m_p}^{K_p} \leqq \delta_{p+1} ;$$

this gives

$$||g_{p+1}||_{m_p}^{K_{p+1}} \leqq M_{p+1}\delta_p + 2\delta_{p+1} \leqq M_{p+1}\delta_p + \delta_p \leqq 2\delta_p M_{p+1} .$$

In particular,

$$||\sum_{q>p} g_q||_{m_p}^{K_{p+1}} \leqq 2 \sum_{q>p} \delta_q M_{q+1} < \tfrac{1}{2}\varepsilon_p .$$

This implies that

$$g = \sum_{q=0}^{\infty} g_q \in C^{m_p}(\Omega) \qquad \text{for all} \quad p.$$

Further, by (1.6.11),

$$\|f-g\|_{m_p}^{L_p} \leqq \|f-\sum_0^p g_q\|_{m_p}^{L_p}+\|\sum_{q>p} g_q\|_{m_p}^{L_p} < \delta_p+\tfrac{1}{2}\varepsilon_p < \varepsilon_p.$$

Thus, if $\{\lambda_p\}$ is a sequence such that $\lambda_p > l_p(\lambda_0, \ldots, \lambda_{p-1})$, the $\{g_p\}$ are defined by the inductive procedure above, and $g = \sum g_p$, then

$$\|f-g\|_{m_p}^{K_{p+1}-K_p} < \varepsilon_p.$$

We shall now show that if the $\lambda_p > l_p(\lambda_0, \ldots, \lambda_{p-1})$ are suitably chosen, then g is analytic.

By definition,

$$\begin{aligned} g_p(x) &= c\lambda_p^{\frac{1}{2}n} \int_{R^n} \varphi_p(y)(f(y)-\sum_{q=0}^{p-1} g_q(y)) \exp(-\lambda_p\|x-y\|^2)dy \\ &= c\lambda_p^{\frac{1}{2}n} \int_{\mathrm{supp}\,(\varphi_p)} \ldots . \end{aligned}$$

Since the integration is over a compact set and $\exp(-\lambda_p\|x-y\|^2)$ is analytic in x, g_p is analytic in R^n for each p. Let now $2\rho_p =$ distance of K_p from $\Omega-K_{p+1}$.

Clearly $\rho_p > 0$. Let U_p be an open set in $C^n \supset R^n$, $U_p \supset K_p$ such that if $z \in U_p$ and $y \in \Omega-K_{p+1}$, we have

$$\mathrm{Re}\,\{(z_1-y_1)^2+ \ldots +(z_n-y_n)^2\} > \rho_p;$$

clearly g_p is the restriction to R^n of the entire function

$$\begin{aligned} h_p(z) = c\lambda_p^{\frac{1}{2}n} \int_{\mathrm{supp}\,(\varphi_p)} & \varphi_p(y)(f(y)-\sum_{q=0}^{p-1} g_q(y)) \\ & \times \exp(-\lambda_p[(z_1-y_1)^2+ \ldots +(z_n-y_n)^2])dy. \end{aligned}$$

Further, since $\mathrm{supp}\,(\varphi_q) \subset \Omega-K_{p+1}$ if $q > p+1$, the integral defining g_q, $q > p+1$, can be replaced by the integral over $\Omega-K_{p+1}$. This shows that, for $z \in U_p$,

1.6.12
$$\begin{aligned} |h_q(z)| &\leqq c\lambda_q^{\frac{1}{2}n} \exp(-\lambda_q\rho_p)H_q(g_0, \ldots, g_{q-1}) \\ &= c\lambda_q^{\frac{1}{2}n}H_q \exp(-\lambda_q\rho_p), \end{aligned}$$

where H_q depends only on $\lambda_0, \ldots, \lambda_{q-1}$, since g_p depends only on the λ_j with $j \leqq p$. We can choose, inductively,

$$\lambda_q > l_q(\lambda_0, \ldots, \lambda_{q-1}),$$

such that

$$\sum \lambda_q^{\frac{1}{2}n} H_q \exp(-\lambda_q \rho) < \infty \qquad \text{for any} \quad \rho > 0.$$

[we have only to choose λ_q such that $\lambda_q^{\frac{1}{2}n} H_q \exp(-\lambda_q/q) < q^{-2}$]. With this choice of $\{\lambda_q\}$, it follows from (1.6.12) that the series

$$h(z) = \sum h_q(z)$$

converges uniformly on U_p for any p; clearly $U = \bigcup U_p$ is an open set in C^n with $U \cap R^n = \Omega$; h is holomorphic on U by Weierstrass' theorem 1.1.3. Its restriction to Ω, which is g, is therefore real analytic in Ω, which proves WHITNEY'S theorem.

§ 1.7 An approximation theorem for holomorphic functions

The question of approximation by holomorphic functions is much more complex than the questions dealt with in § 1.6. In the first place, since the uniform limit of holomorphic functions is again holomorphic, we can at best hope to approximate *holomorphic* functions by polynomials in the complex variables $z_1, \ldots, z_n$. There are, however, geometric and analytic conditions on an open set $U \subset C^n$ in order that any holomorphic function on U be approximable by polynomials.

The simplest example of a domain where approximation is not possible is $C^* = \{z \in C | z \neq 0\}$; the function z^{-1} cannot be approximated by polynomials in z. In the case of one variable, the restrictions are *topological* (see theorems 1.7.2 and 3.10.11) but for $n > 1$, it is known that this is no longer the case.

1.7.1 DEFINITION. An open set $U \subset C^n$ is called a Runge domain if every holomorphic function on U can be approximated by polynomials in $z_1, \ldots, z_n$, uniformly on compact subsets of U.

The theorem that follows is contained in the general theorem proved in chap. 3 (see § 3.10). For a simple direct proof based on Cauchy's integral formula, see e.g. HÖRMANDER [1966].

1.7.2 RUNGE'S THEOREM. An open connected set $U \subset C$ is a Runge domain if and only if every connected component of U is simply connected.

Let now U be an open set in C^n and $\lambda: U \to R^+$ a continuous function with $\lambda(z) > 0$ for all $z \in U$. Let $dv = dv_z$ denote Lebesgue measure in C^n (with variables $z_1, \ldots, z_n$). Let $\mathscr{H}(\lambda)$ be the set of holomorphic functions on U such that

$$\int_U |f(z)|^2 \lambda(z) dv < \infty$$

1.7.3 LEMMA. If $f, g \in \mathscr{H}(\lambda)$, set

$$(f, g)_\lambda = (f, g) = \int_U f(z)\overline{g(z)}\lambda(z) dv.$$

With the scalar procuct (f, g), $\mathscr{H}(\lambda)$ is a Hilbert space.

PROOF. Since the space $L^2(\lambda dv)$ of square integrable functions with respect to the measure $\lambda(z)dv_z$ is complete, we have only to show that $\mathscr{H}(\lambda)$ is closed in $L^2(\lambda dv)$. This results at once from the following proposition.

1.7.4 PROPOSITION. If $\{f_p\}$ is a sequence of elements of $\mathscr{H}(\lambda)$ and

$$\int_U |f_p(z) - f_q(z)|^2 \lambda(z) dv_z \to 0 \quad \text{as} \quad p, q \to \infty,$$

then $\{f_p\}$ converges uniformly on compact sets in U.

PROOF. Since λ is bounded below on any compact subset of U by a positive constant, we may suppose that $\lambda \equiv 1$.

If g is holomorphic in a neighbourhood of the closed disc $\{z \in C \mid |z-a| \leqq \rho\}$, we have, by Cauchy's formula,

$$g(a) = (\pi\rho^2)^{-1} \int_{|z| \leqq \rho} g(a+z) dv.$$

Applying this n times, it follows that if $h(z_1, \ldots, z_n)$ is holomorphic in a neighbourhood of the polycylinder $|z_1 - a_1| \leqq \rho, \ldots, |z_n - a_n| \leqq \rho$, then

$$h(a) = (\pi\rho^2)^{-n} \int_{|z| \leqq \rho} h(a+z) dv.$$

Let K be a compact set in U and $\rho > 0$ be so small that the set

$$K_\rho = \{z \in C^n | \text{ there is an } a \in K \quad \text{with} \quad |z-a| \leqq \rho\}$$

is compact in U. If f is holomorphic in U and $a \in K$, we have

$$|f(a)| \leqq (\pi\rho^2)^{-n} \int_{|z| \leqq \rho} |f(a+z)| dv \leqq (\pi\rho^2)^{-n} \int_{K_\rho} |f(z)| dv.$$

If we apply this inequality to the squared differences $(f_p - f_q)^2$, we obtain the proposition.

Let $\{\varphi_\nu\}$ be a complete orthonormal system in $\mathscr{H}(\lambda)$. Then if $f \in \mathscr{H}(\lambda)$, we have

$$f = \sum c_\nu \varphi_\nu, \qquad c_\nu = (f, \varphi_\nu),$$

the series being convergent in $\mathscr{H}(\lambda)$. From theorem 1.7.4, we deduce:

1.7.5 Lemma. If $\{\varphi_\nu\}$ is a complete orthonormal system in $\mathscr{H}(\lambda)$, then any $f \in \mathscr{H}(\lambda)$ can be approximated, uniformly on compact subsets of U, by finite linear combinations

$$\sum_{\nu=1}^{N} c_\nu \varphi_\nu, \qquad c_\nu \in C.$$

1.7.6 Proposition. Let U_1, U_2 be open sets in C^n, C^m respectively and let λ_j be a strictly positive continuous function on U_j. Define $\lambda_1 \times \lambda_2$ on $U_1 \times U_2$ by

$$(\lambda_1 \times \lambda_2)(z_1, z_2) = \lambda_1(z_1)\lambda_2(z_2).$$

Let $\{\varphi_\nu^{(j)}\}$ be a complete orthonormal system in $\mathscr{H}(\lambda_j)(j = 1, 2)$. Then the functions $\{\varphi_{\nu_1}^{(1)}(z_1)\varphi_{\nu_2}^{(2)}(z_2)\}$ form a complete orthonormal system in $\mathscr{H}(\lambda_1 \times \lambda_2)$.

Proof. It is sufficient to show that if $f \in \mathscr{H}(\lambda_1 \times \lambda_2)$ and

$$\int_{U_1 \times U_2} f(z_1, z_2)\overline{\varphi_{\nu_1}^{(1)}(z_1)}\, \overline{\varphi_{\nu_2}^{(2)}(z_2)}\lambda_1(z_1)\lambda_2(z_2) dv = 0 \qquad \text{for all} \quad \nu_1, \nu_2$$

(where dv is Lebesgue measure in $C^{n_1+n_2}$), then $f \equiv 0$. Let dv_j be Lebesgue measure in C^{n_j}. We first show that for $a_1 \in U_1$, the function

$$z_2 \mapsto f(a_1, z_2)$$

on U_2 belongs to $\mathscr{H}(\lambda_2)$. In fact, it follows from the proof of proposition 1.7.4 that for $z_2 \in U_2$, and $\rho > 0$ small enough,

$$|f(a_1, z_2)|^2 \leqq c^{-1}(\pi\rho^2)^{-n_1} \int_{|z_1 - a_1| \leqq \rho} |f(z_1, z_2)|^2 \lambda_1(z_1) dv_1,$$

where

$$c = \inf \lambda_1(z_1) \quad \text{for} \quad |z_1 - a_1| \leqq \rho,$$

so that

$$\int_{U_2} |f(a_1, z_2)|^2 \lambda_2(z_2) dv_2$$
$$\leqq c^{-1}(\pi\rho^2)^{-n_1} \int_{U_1 \times U_2} |f(z_1, z_2)|^2 \lambda_1(z_1)\lambda_2(z_2) dv < \infty.$$

We claim now that for any ν_2, the function

$$g(z_1) = g^{(\nu_2)}(z_1) = \int_{U_2} f(z_1, z_2)\overline{\varphi_{\nu_2}^{(2)}(z_2)}\lambda_2(z_2) dv_2$$

(which is well-defined by what we have seen above) lies in $\mathscr{H}(\lambda_1)$. In the first place $g(z_1)$ is holomorphic in U_1 since, if $\{K_p\}$ is a sequence of compact sets exhausting U_2,

$$\int_{K_p} f(z_1, z_2)\overline{\varphi_{\nu_2}^{(2)}(z_2)}\lambda_2(z_2) dv_2$$

converges to $g(z_1)$, uniformly on compact subsets of U_1, as in proposition 1.7.4. Further, by Schwarz' inequality

$$|g(z_1)|^2 \leqq \int_{U_2} |f(z_1, z_2)|^2 \lambda_2(z_2) dv_2 \int_{U_2} |\varphi_{\nu_2}^{(2)}(z_2)|^2 \lambda_2(z_2) dv_2,$$

so that

$$\int_{U_1} |g(z_1)|^2 \lambda_1(z_1) dv_1 \leqq \int_{U_1} \lambda_1(z_1) dv_1 \int_{U_2} |f(z_1, z_2)|^2 \lambda_2(z_2) dv_2 < \infty,$$

since $f \in \mathscr{H}(\lambda_1 \times \lambda_2)$. Thus $g(z_1) \in \mathscr{H}(\lambda_1)$. By hypothesis, $g(z_1)$ is orthogonal to all the $\{\varphi_{\nu_1}^{(1)}(z_1)\}$ in $\mathscr{H}(\lambda_1)$, hence $g(z_1) \equiv 0$. Thus, for fixed z_1, $f(z_1, z_2)$ is orthogonal to all the $\{\varphi_{\nu_2}^{(2)}(z_2)\}$ in $\mathscr{H}(\lambda_2)$, so that $f(z_1, z_2) \equiv 0$.

1.7.7 Theorem. If U_1, U_2 are open sets in C^{n_1}, C^{n_2} respectively, the finite linear combinations

$$\sum_\nu \varphi_\nu^{(1)}(z_1)\varphi_\nu^{(2)}(z_2),$$

where $\varphi_\nu^{(j)}(z_j)$ is holomorphic on U_j, are dense in the space of holomorphic functions on $U_1 \times U_2$ with respect to the topology of compact convergence.

Proof. Let $f(z_1, z_2)$ be holomorphic on $U_1 \times U_2$. There exists a strictly positive continuous function $\eta: U_1 \times U_2 \to R^+$ such that $f \in \mathscr{H}(\eta)$, i.e.,

$$\int_{U_1 \times U_2} |f|^2 \eta \, dv < \infty.$$

Let $\{K_p^{(j)}\}$ be compact sets in U_j such that

$$K_p^{(j)} \subset \mathring{K}_{p+1}^{(j)}, \qquad \bigcup K_p^{(j)} = U_j.$$

Then

$$\bigcup_p K_p^{(1)} \times K_p^{(2)} = U_1 \times U_2.$$

Let $0 < \varepsilon_p < 1$ be such that $\eta(z_1, z_2) \geqq \varepsilon_p$ for $(z_1, z_2) \in K_p^{(1)} \times K_p^{(2)}$. Let λ_j be a strictly positive continuous function on U_j such that $\lambda_j(z_j) \leqq \varepsilon_p$ for $z_j \in K_p^{(j)} - K_{p-1}^{(j)}$ [which exists; see remark 1.6.6]. Now

$$K_p^{(1)} \times K_p^{(2)} - K_{p-1}^{(1)} \times K_{p-1}^{(2)} = K_p^1 \times (K_p^{(2)} - K_{p-1}^{(2)}) \cup (K_p^{(1)} - K_{p-1}^{(1)}) \times K_p^{(2)}.$$

It follows trivially that for

$$(z_1, z_2) \in K_p^{(1)} \times K_p^{(2)} - K_{p-1}^{(1)} \times K_{p-1}^{(2)},$$

we have

$$\lambda_1(z_1)\lambda_2(z_2) \leqq \varepsilon_p \leqq \eta(z_1, z_2).$$

Hence we have

$$\lambda_1(z_1)\lambda_2(z_2) \leqq \eta(z_1, z_2) \quad \text{on} \quad U_1 \times U_2.$$

Hence $f \in \mathscr{H}(\lambda_1 \times \lambda_2)$.

If now $\{\varphi_\nu^{(j)}\}$ is a complete orthonormal system in $\mathscr{H}(\lambda_j)$, the products $\varphi_{\nu_1}^{(1)}(z_1)\varphi_{\nu_2}^{(2)}(z_2)$ form a complete orthonormal system in $\mathscr{H}(\lambda_1 \times \lambda_2)$. Since $f \in \mathscr{H}(\lambda_1 \times \lambda_2)$, by lemma 1.7.5, there are complex constants $c_{\nu_1\nu_2}$ so that finite linear combinations of the form

$$\sum c_{\nu_1\nu_2}\varphi_{\nu_1}^{(1)}(z_1)\varphi_{\nu_2}^{(2)}(z_2)$$

approximate f uniformly on ompact subsets of $U_1 \times U_2$.

1.7.8 COROLLARY. If U_j is a Runge domain in $C^{n_j}(j = 1, 2)$, then $U_1 \times U_2$ is a Runge domain in $C^{n_1+n_2}$. In particular, if $U_1, \ldots, U_n$ are simply connected open sets in C, then $U_1 \times \ldots \times U_n$ is a Runge domain in C^n.

We shall take up deeper properties of Runge domains in C^n later.

§ 1.8 Ordinary differential equations

1.8.1 LEMMA. Let I be an interval containing 0 in R and let $w: I \to R^+$ be a continuous function. Let $M, \eta \in R$, $M > 0$, $\eta \geqq 0$, and suppose that for $t \in I$ we have

1.8.2 $w(t) \leqq \varepsilon M \int_0^t w(s)ds + \eta$, where $\varepsilon = +1$ if $t \geqq 0$, $\varepsilon = -1$ if $b < 0$.

Then we have also $w(t) \leqq \eta e^{M|t|}$ for $t \in I$.

PROOF. Let first $t \geqq 0$. We have

$$e^{Mt}\frac{d}{dt}\left\{e^{-Mt}\int_0^t w(s)ds\right\} = w(t) - M\int_0^t w(s)ds \leqq \eta,$$

so that

$$e^{-Mt}\int_0^t w(s)ds \leqq \eta\int_0^t e^{-Ms}ds = \frac{\eta}{M}(1-e^{-Mt}).$$

Combining this with (1.8.2), we obtain

$$w(t) \leqq \eta e^{Mt} \qquad (\text{since } M > 0).$$

If $t < 0$, let $\tau = -t > 0$. We have

$$w(-\tau) \leqq M\int_{-\tau}^{0} w(s)\mathrm{d}s + \eta = M\int_{0}^{\tau} w(-s)\mathrm{d}s + \eta,$$

and by what we have seen,

$$w(-\tau) \leqq \eta e^{M\tau},$$

i.e.,

$$w(t) \leqq \eta e^{M|t|},$$

also for $t < 0$.

1.8.3 Definition. Let Ω, Ω' be subsets of R^n, R^m respectively, and let $f\colon \Omega \times \Omega' \to R^p$ be a map. We say that f satisfies a Lipschitz condition in $x \in \Omega$ on a set $S \times S' \subset \Omega \times \Omega'$ ($S \subset \Omega$, $S' \subset \Omega'$) uniformly in $x' \in S'$, if there is $M > 0$ such that

$$\|f(x, x') - f(y, x')\| \leqq M\|x - y\|$$

for $(x, x'), (y, x') \in S \times S'$.

1.8.4 Theorem. Let Ω, Ω' be open sets in R^n, R^m respectively and I an open interval in R with $0 \in I$. Let

$$f\colon \Omega \times I \times \Omega' \to R^n$$

be a continuous map; we denote a point in $\Omega \times I \times \Omega'$ by (x, t, α). Suppose that for any compact sets $K \subset \Omega$, $K' \subset \Omega'$, f satisfies a Lipschitz condition in x on $K \times I \times K'$, uniformly in t, α.

Then, given $x_0 \in \Omega$ and a compact set $K' \subset \Omega'$, there is an interval $I_0 = \{t \mid |t| < \varepsilon\}$ and, for each $\alpha \in K'$ a unique C^1 map $I_0 \to \Omega$, $t \mapsto x(t, \alpha)$ such that

1.8.5 $$f(x(t, \alpha), t, \alpha) = \frac{\partial x}{\partial t}(t, \alpha), \qquad x(0, \alpha) = x_0.$$

Moreover, the map $I_0 \times K' \to \Omega$ given by $(t, \alpha) \mapsto x(t, \alpha)$ is continuous.

PROOF. Let $M > 0$ be such that for $x, y \in K$, $\alpha \in K'$ we have

$$\|f(x, t, \alpha) - f(y, t, \alpha)\| \leqq M\|x-y\|.$$

Let $r > 0$ be such that

$$\Omega_0 = \{x \mid \|x - x_0\| \leqq r\} \subset \Omega;$$

let K be such that $\Omega_0 \subset K$. Let $\|f\| < C$ on $\Omega_0 \times I \times K'$. Let $\varepsilon' > 0$ be such that $\{t \mid |t| \leqq \varepsilon'\} \subset I$ and let

$$I_0 = \{t \mid |t| < \varepsilon\},$$

where $\varepsilon = \min(\varepsilon', r/C)$. Define $x_0(t, \alpha) \equiv x_0$ and, for $n > 0$, $x_n\colon I_0 \times K' \to R^n$ by

1.8.6 $$x_n(t, \alpha) = x_0 + \int_0^t f(x_{n-1}(s, \alpha), s, \alpha)ds$$

We claim that $x_n(t, \alpha) \in \Omega_0$ for $(t, \alpha) \in I_0 \times K'$. In fact, this is trivial for $n = 0$. If this is already proved for x_{n-1}, we have

$$\left\| \int_0^t f(x_{n-1}(s, \alpha), s, \alpha)ds \right\| \leqq C|t| < C\varepsilon \leqq r,$$

so that

$$\|x_n(t, \alpha) - x_0\| < r.$$

Moreover, we have

$$\|x_{n+1}(t, \alpha) - x_n(t, \alpha)\| \leqq \frac{1}{n!} M^n C |t|^n,$$

for $n \geqq 0$. In fact,

$$\|x_1 - x_0\| = \left\| \int_0^t f(x_0, s, \alpha)ds \right\| \leqq |t|C,$$

which is our inequality for $n = 0$. If we have the inequality for $n = m$, then

$$\begin{aligned}
&\|x_{m+2}(t, \alpha) - x_{m+1}(t, \alpha)\| \\
&\quad = \left\| \int_0^t \{f(x_{m+1}(s, \alpha), s, \alpha) - f(x_m(s, \alpha), s, \alpha)\}ds \right\| \\
&\quad \leqq M \frac{1}{m!} M^m C \int_0^{|t|} s^m ds = \frac{1}{(m+1)!} M^{m+1} C |t|^{m+1},
\end{aligned}$$

which is what we want. Hence, as $n \to \infty$, $x_n(t, \alpha)$ converges uniformly to a continuous function $x(t, \alpha)$. Moreover, letting $n \to \infty$ in (1.8.6), we have

$$x(t, \alpha) = x_0 + \int_0^t f(x(s, \alpha), s, \alpha)ds,$$

which implies that, for fixed α, the map $t \mapsto x(t, \alpha)$ is C^1.

Finally, if for some $\alpha_0 \in K'$, $u: I_0 \to \Omega$ is a C^1 map with

$$f(u(t), t, \alpha_0) = \frac{du}{dt}(t), \qquad u(0) = x_0,$$

let

$$w(t) = x(t, \alpha_0) - u(t).$$

Then we have, for $t \geqq 0$,

$$||w(t)|| \leqq M \int_0^t ||w(s)||ds,$$

and, by lemma 1.8.1 with $\eta = 0$, this implies that $w(t) = 0$ for $t \geqq 0$. A similar argument applies if $t < 0$, which gives us the uniqueness statement.

1.8.7 REMARK. If $\Omega = R^n$ and f satisfies an estimate

$$||f(x, t, \alpha)|| \leqq C_1||x|| + C_2, \qquad C_1, C_2 > 0$$

on $R^n \times I \times K'$, then x is defined (and unique) on $I \times \Omega'$. In fact, if we define x_n by (1.8.6) for any fixed α and $t \in I$, we obtain, for $t \geqq 0$

$$||x_n(t, \alpha)|| \leqq M_1 \int_0^t ||x_{n-1}(s)||ds + M_2,$$

and we can suppose that

$$||x_0|| \leqq M_2 e^{M_1 t}.$$

It follows, by induction, that

$$||x_n(t, \alpha)|| \leqq M_2 e^{M_1 t},$$

so that the sequence $\{x_n(t, \alpha)\}_{n \geqq 0}$ is *uniformly bounded* ($t \geqq 0$;

similar arguments show that the result is also true for $t < 0$). Now we can apply the Lipschitz condition on f to show that

$$||x_n(t, \alpha) - x_{n+1}(t, \alpha)|| \leqq \frac{1}{n!} AM^n |t|^n, \qquad n \geqq 0,$$

so that we can repeat the proof of theorem 1.8.4. In particular, if f is *linear in x*, the solution of (1.8.5) exists on $I \times \Omega'$.

1.8.8 THEOREM. Let the notation be as in theorem 1.8.4 and J be an open interval containing the closure of I. Suppose $f \in C^k(\Omega \times J \times \Omega')$, $k \geqq 1$ (in particular, f satisfies a Lipschitz condition as in theorem 1.8.4). Then the solution x of (1.8.5) belongs to $C^k(I_0 \times \mathring{K}')$.

PROOF. Let $U' = \mathring{K}'$. We first show that if $f \in C^1(\Omega \times J \times \Omega')$, then $x^1 \in C^1(I_0 \times U')$. If $\alpha = (\alpha_1 \ldots, \alpha_m)$, we have only to show that

$$\frac{\partial x}{\partial \alpha_j}, \qquad j = 1, \ldots, m$$

exists and is continous on $I_0 \times U'$, since, by (1.8.5), $\partial x / \partial t$ exists and is continuous. We shall suppose that $t \geqq 0$. For fixed t, α, let $h_{t,\alpha}: \Omega \to R^n$ be the map $x \mapsto f(x, t, \alpha)$ and let

$$A(t, \alpha) = (\mathrm{d}h)(x(t, \alpha)) = (\mathrm{d}_1 f)(x(t, \alpha), t, \alpha);$$

$A(t, \alpha)$ is a linear transformation of R^n into itself. Let

$$B(t, \alpha) = \frac{\partial f}{\partial \alpha_j}(x(t, \alpha), t, \alpha);$$

B is a continuous map of $I_0 \times U'$ into R^n.

By the remark 1.8.7, there is a continuous map

$$y: I_0 \times U' \to R^n,$$

(which, for fixed α, is C^1) such that

1.8.9
$$\frac{\partial y}{\partial t} = A(t, \alpha) y + B(t, \alpha), \qquad y(0, \alpha) = 0.$$

For fixed $\alpha = (\alpha_1, \ldots, \alpha_m) \in U'$ and small enough real $h \neq 0$, we set

$$\alpha^h = (\alpha_1, \ldots, \alpha_{j-1}, \alpha_j + h, \alpha_{j+1}, \ldots, \alpha_m)$$

and

$$u_h(t) = h^{-1}(x(t, \alpha^h) - x(t, \alpha)).$$

By Taylor's formula, we have, for $0 \leqq s \leqq t$,

$$f(x(s, \alpha^h), s, \alpha^h) - f(x(s, \alpha), s, \alpha) = hA(s, \alpha)u_h(s) + hB(s, \alpha) + \varepsilon(s, h),$$

where

$$\varepsilon(s, h) = o(|h| \; \|u_h(s)\| + |h|),$$

uniformly in s as $h \to 0$. (Note that $|h| \; \|u_h(s)\| + |h| = \|x(s, \alpha^h) - x(s, \alpha)\| + \|\alpha^h - \alpha\|$). Hence

$$u_h(t) = \int_0^t \{A(s, \alpha)u_h(s) + B(s, \alpha) + \delta(s, h)\} ds,$$

where

$$\delta(s, h) = \frac{1}{h} \varepsilon(s, h) = o(\|u_h(s)\| + 1).$$

This implies that there is a constant $C_1 > 0$ such that

$$\|u_h(t)\| \leqq C_1 \int_0^t \|u_h(s)\| ds + C_1,$$

so that, by lemma 1.8.1, $\|u_h(t)\|$ is bounded, uniformly in t, as $h \to 0$. Hence $\delta(s, h) \to 0$ uniformly in s as $h \to 0$.

Let now

$$z_h(t) = u_h(t) - y(t, \alpha).$$

Then, since

$$y(t, \alpha) = \int_0^t \{A(s, \alpha)y(s, \alpha) + B(s, \alpha)\} ds,$$

we have

$$z_h(t) = \int_0^t A(s, \alpha) z_h(s) ds + \int_0^t \delta(s, h) ds.$$

Since

$$\eta = \left\| \int_0^t \delta(s, h) ds \right\| \to 0$$

as $h \to 0$, lemma 1.8.1 shows that $z_h(t) \to 0$ as $h \to 0$. This means precisely that $(\partial x/\partial \alpha_j)(\alpha)$ exists and equals $y(t, \alpha)$. Hence, as remarked earlier, $x \in C^1(I_0 \times U')$.

To finish the proof of theorem (1.8.8), we now proceed by induction on k. If $f \in C^k(\Omega \times J \times \Omega')$, $k > 1$, and if $x \in C^{k-1}(I_0 \times U')$, then, clearly, $A(t, \alpha)$, $B(t, \alpha)$ are C^{k-1} functions. Since y satisfies the equation

$$\frac{\partial y}{\partial t} = A(t, \alpha)y + B(t, \alpha),$$

the induction assumption implies that $y \in C^{k-1}(I_0 \times U')$. Since we have

$$\frac{\partial x}{\partial \alpha_j}(t, \alpha) = y(t, \alpha), \qquad \frac{\partial x}{\partial t} = f(x(t, \alpha), t, \alpha),$$

all first derivatives of x are C^{k-1}, so that $x \in C^k(I_0 \times U')$.

These results have obvious analogues for holomorphic functions.

1.8.10 THEOREM. Let U, U' be open sets in C^n, C^m respectively and $D = \{z | \, |z| < \rho\}$ a disc in the complex plane. If f: $U \times D \times U' \to C^n$ is a holomorphic map, then, for any $x_0 \in U$, and compact $K' \subset U'$, there is a $\delta > 0$ such that for each $\alpha \in K'$, there is a unique holomorphic map

$$x: z \mapsto x(z, \alpha)$$

of $D_\delta \to \Omega$, $\{D_\delta = \{z | \, |z| < \delta\}$ for which

$$\frac{\partial x}{\partial z} = f(x(z, \alpha), z, \alpha), \qquad x(0, \alpha) = x_0 .$$

Further, the map $(z, \alpha) \mapsto x(z, \alpha)$ from $D_\delta \times \mathring{K}'$ into Ω is holomorphic.

PROOF. The proof is the same as in theorem 1.8.4; we define

$$x_n(z, \alpha) = x_0 + \int_0^z f(x_{n-1}(\zeta, \alpha), \zeta, \alpha) d\zeta,$$

the integration being along the line segment from 0 to z. Noting that there is automatically a Lipschitz condition, we prove that

$$\|x_{n+1}(z, \alpha) - x_n(z, \alpha)\| \leqq \frac{1}{n!} M^n C|z|^n;$$

clearly each x_n is holomorphic; hence so is $\lim_{n\to\infty} x_n$. The uniqueness is proved as before.

1.8.11 COROLLARY. If in theorem 1.8.4, f is a real analytic map then the solution x of (1.8.5) is real analytic in $J_0 \times U'$, where J_0 is some open set interval containing the origin.

PROOF. We can find open sets W, D, W' in C^n, C, C^m respectively with $W \cap R^n = \Omega$, $D \cap R = I$, $W' \cap R^m = \Omega'$ such that f admits an extension to $W \times D \times W'$ as a holomorphic map (which we again denote by f). The solution w of the equation

$$\frac{\partial w}{\partial z} = f(w(z, \alpha), z, \alpha), \qquad w(0) = x_0$$

is then defined and holomorphic in a neighbourhood of $J_0 \times U'$ where J_0 is an interval on the real axis with $0 \in J_0$. Moreover, w is the limit of w_n defined inductively by

$$w_0 \equiv x_0, \qquad w_n(z, \alpha) = x_0 + \int_0^z f(w_{n-1}(\zeta, \alpha), \zeta, \alpha)\mathrm{d}\zeta$$

and so is clearly real for real z, α. Hence, its restriction to $J_0 \times U'$, which is the solution we are seeking, is analytic.

1.8.12 THEOREM. Let Ω, I, Ω' be as in theorem 1.8.4, and let $f\colon \Omega \times I \times \Omega' \to R^n$ be a C^k (resp. real analytic) map, $1 \leqq k \leqq \infty$. Then, for any point $(u_0, u_0, \alpha_0, \xi_0) \in I \times I \times \Omega' \times \Omega$, there is a neighbourhood $W = J \times J \times U' \times U$ and a C^k (resp. real analytic) map $x\colon W \to R^n$ such that

$$x(u, u, \alpha, \xi) = \xi, \qquad \frac{\partial x}{\partial t}(t, u, \alpha, \xi) = f(x(t, u, \alpha, \xi), t, \alpha),$$

for $t, u \in J$, $\alpha \in U'$, $\xi \in U$.

In other words, in the neighbourhood of a given point $(u_0, u_0, \alpha_0, \xi_0)$, the solution of (1.8.5) which takes the value ξ at the point u, depends differentiably (resp. analytically) on t, u, ξ as well as on any parameters in f.

PROOF. We consider a neighbourhood Δ of the point $(0, 0, u_0, \alpha_0, \xi_0) \in R^n \times R \times I \times \Omega' \times \Omega$ and a C^k (resp. analytic) map $g: \Delta \to R^n$ given by

$$g(y, t, u, \alpha, \xi) = f(\xi + y, t + u, \alpha).$$

If $y(t, u, \alpha, \xi)$ is the solution of

$$\frac{\partial y}{\partial t} = g(y, t, u, \alpha, \xi), \quad \text{with} \quad y(0, u, \alpha, \xi) = 0,$$

then we have

$$x(t, u, \alpha, \xi) = \xi + y(t - u, u, \alpha, \xi),$$

and the result follows from theorem 1.8.8 and corollary 1.8.11.

1.8.13 REMARK. The above theorem has an analogue for holomorphic functions which we do not formulate.

Finally, we remark that theorems 1.8.8 and 1.8.12 can be strengthened to the following statement.

1.8.14 THEOREM. Let Ω, I, Ω' be as in theorem 1.8.3, $f: \Omega \times I \times \Omega' \to R^n$ be a C^k (resp. analytic) map and $I_1 \subset I$, $\Omega'_1 \subset \Omega'$ be connected open subsets. Then if $x: I_1 \times \Omega'_1 \to \Omega$ is a continuous map which is C^1 in t for fixed α and

$$\frac{\partial x(t, \alpha)}{\partial t} = f(x(t, \alpha), t, \alpha)$$

and if $x(t_0, \alpha)$ is C^k resp. analytic for some $t_0 \in I_1$, then $x: I_1 \times \Omega'_1$ is C^k (resp. analytic). The proof uses theorem 1.8.12. We omit the details.

CHAPTER 2

Manifolds

§ 2.1 Basic definitions

2.1.1 DEFINITION. Let V be a Hausdorff topological space. We say that V is a *manifold* (or a C^0 manifold or a locally Euclidean space) *of dimension n* if every point $a \in V$ has an open neighbourhood, which is homeomorphic to an open set in R^n.

2.1.2 DEFINITION. Let V be as above and let $0 \leqq k \leqq \infty$. We call V a C^k *manifold* or a (differentiable) manifold of class C^k and dimension n if there is given a family of pairs (U_i, φ_i), i running over an indexing set $\mathscr{I}$, where U_i is an open set in V and φ_i is a homeomorphism of U_i onto an open set in R^n such that

(a) $$\bigcup_{i \in \mathscr{I}} U_i = V$$

and (b) whenever $i, j \in \mathscr{I}$ are such that $U_i \cap U_j \neq \emptyset$, the map

$$\varphi_j \circ \varphi_i^{-1}\colon \varphi_i(U_i \cap U_j) \to \varphi_j(U_i \cap U_j) \text{ is a } C^k \text{ map.}$$

If the $\{(U_i, \varphi_i)\}_{i \in \mathscr{I}}$ can be so chosen that whenever $U_i \cap U_j \neq \emptyset$, the map $\varphi_j \circ \varphi_i^{-1}\colon \varphi_i(U_i \cap U_j) \to R^n$ is actually real analytic, we say that V is a *real analytic manifold.*

We write

$$\dim V = \dim_R V$$

for the dimension of a manifold (see remark 2.1.4).

2.1.3 DEFINITION. If V is a manifold, a C^k structure on V is a *maximal set* $\mathfrak{B} = \{(U_i, \varphi_i)\}_{i \in \mathscr{I}}$ of pairs with the properties (a) and (b) of definition 2.1.2. Elements of $\mathfrak{B}$ are called *coordinate systems* of this C^k structure. If (U, φ) is a coordinate system, U is called a *coordinate neighbourhood* and $\varphi = (\varphi_1, \ldots, \varphi_n)$ (often written $x_1, \ldots, x_n$) coordinates in U. The mappings $\varphi_j \circ \varphi_i^{-1}$ above are called *coordinate transformations*.

Note that any system of pairs as in definition 2.1.2 can be completed to a unique C^k structure; two such systems are equivalent if they are contained in the same C^k structure on V. We can thus speak of coordinate systems and so on on a C^k manifold. An open subset of a C^k manifold carries a natural induced C^k structure.

2.1.4 REMARK. The dimension of a manifold is an invariant of the manifold (independent of the local homeomorphisms used). This follows from a theorem of Brouwer which asserts that a non-empty open set in R^n is homeomorphic to one in R^m only if $m = n$. For a proof, see e.g. HUREWICZ and WALLMAN [1948]. The corresponding invariance statement for C^k manifolds, $k \geqq 1$, is much simpler and will be proved later.

We note that it follows from a theorem of DIEUDONNÉ [1944] (see also BOURBAKI [1965]) that, for a manifold V, the following conditions are equivalent.

1. V is paracompact; i.e., any open covering of V has a locally finite refinement.
2. Every connected component of V is a countable union of compact sets.
3. Every connected component of V has a countable base for its open sets.

2.1.5 DEFINITION. A Hausdorff topological space V is called a *complex manifold of complex dimension n* if there is given a family $\{(U_i, \varphi_i)\}_{i \in \mathscr{I}}$, where φ_i is a homeomorphism of U_i onto an open set in C^n and $\varphi_j \circ \varphi_i^{-1}$ is holomorphic on $\varphi_i(U_i \cap U_j)$. We define a *complex analytic* (or just complex) structure as in definition 2.1.3. We write $n = \dim V = \dim_C V$ for the dimension of V.

2.1.6 DEFINITION. If V is a C^k manifold, U an open set in V, a map $f: U \to R$ is called a *C^r function* on U if, for any coordinate system (W, ψ) on V with $W \subset U$, the function

$$f \circ \psi^{-1}: \psi(W) \to R$$

is C^r $(0 \leqq r \leqq k)$. The set of C^r functions on V is denoted by $C^r(V)$. The *support* supp (f) of a C^r function f on V is, again, the closure in V of the set $\{x \in V | f(x) \neq 0\}$. The set of $f \in C^r(V)$ for which supp (f) is compact is denoted by $C_0^r(V)$.

Let V, V' be C^k manifolds, $\{(U_i, \varphi_i)\}_{i \in \mathscr{I}}$, $\{(U_j', \varphi_j')\}_{j \in \mathscr{J}}$ their defining C^k structures. A continuous map $f: V \to V'$ is called a *C^r map* $(0 \leqq r \leqq k)$ if, for any pair of coordinate systems (U_i, φ_i) on V, (U_j', φ_j') on V' such that $f(U_i) \subset U_j'$, the map $\varphi_j' \circ f \circ \varphi_i^{-1}: \varphi_i(U_i) \to \varphi_j'(U_j')$ is a C^r map. The set of C^r maps of V into V' is denoted by $C^r(V, V')$.

Real analytic and holomorphic functions and mappings between real analytic and complex manifolds are defined similarly.

Let V, V' be C^k (real, complex analytic) manifolds. A continuous map $f: V \to V'$ is C^k (real analytic, holomorphic) if and only if the following condition is satisfied.

For any open set U' in V' and any C^k (real analytic, holomorphic) function g' on U', $g' \circ f$ is a C^k (real analytic, holomorphic) function on $f^{-1}(U')$.

If V, V' are C^k manifolds and $f: V \to V'$ a homeomorphism such that f and f^{-1} are C^k mappings, we call f a C^k diffeomorphism (or diffeomorphism or C^k isomorphism) between V and V'. V, V' are called diffeomorphic (C^k diffeomorphic, C^k isomorphic) if there is a C^k diffeomorphism $f: V \to V'$. Real analytic and holomorphic (= complex analytic) isomorphisms between corresponding manifolds are similarly defined.

2.1.7 EXAMPLES.

(a) $S^1 = \{x \in R^2 | \, ||x|| = 1\}$ is a 1-dimensional C^∞ manifold.

(b) Let V be a C^k manifold and $\tilde{V}$ a Hausdorff space. Let $p: \tilde{V} \to V$ be a local homeomorphism, i.e., any $a \in \tilde{V}$ has a neighbourhood U such that $p(U)$ is open in V and $p: U \to p(U)$ is a homeomorphism.

Then there is a unique C^k structure on $\tilde{V}$ for which p is a local C^k diffeomorphism (i.e., for any $a \in \tilde{V}$, there is a neighbourhood U such that $p|U: U \to p(U)$ is a C^k diffeomorphism; note that $p(U)$ is open in V). A similar remark applies to real analytic and complex manifolds.

(c) If V, W are C^k manifolds, $V \times W$ carries a natural structure of a C^k manifold for which the projections are C^k maps.

It is clear that a complex analytic manifold carries a natural real analytic structure, a real analytic manifold a C^∞ structure and a C^k manifold ($0 \leqq k \leqq \infty$) a C^r structure if $0 \leqq r \leqq k$. Conversely, it follows from results of WHITNEY [1936] that any paracompact C^r manifold, $r \geqq 1$, carries a real analytic structure compatible with the given C^r structure. Further, the imbedding theorem of GRAUERT [1958] (see § 2.15 for the statement) and the approximation theorem of Whitney 1.6.5 imply that this structure is unique (upto isomorphism; the identity map need not be an isomorphism). It may happen that a C^0 manifold carries no C^1 structure (KERVAIRE [1960]) and even if it does, this structure may not be unique. For example, MILNOR [1956] has shown that the sphere S^7 (see example 2.5.6) can carry, besides its natural structure, a C^∞ structure such that there is *no* C^1 diffeomorphism between the two (not only that the identity is not a diffeomorphism).

The problem of the existence and uniqueness of complex structures is a problem of a completely different nature and has given rise to a vast literature. (See in particular HOPF [1948], KODAIRA and SPENCER [1958].)

Since the results of Milnor and Kervaire, much more information has been obtained concerning the existence and uniqueness of differential structures on topological manifolds. Several papers dealing with this problem will be found in the Proceedings of the International Mathematical Congress of 1962 and 1966.

Let V be a C^k manifold and $a \in V$. Consider all pairs (f, U), where U is an open set containing a and $f \in C^k(U)$. We say that two such pairs are equivalent, $(f, U) \sim (f', U')$ if there is an open set $W \subset U \cap U'$, $a \in W$, such that $f|W = f'|W$. This is clearly an equivalence relation. An equivalence class is called a *germ* of C^k functions at a.

We shall frequently identify a germ with a C^k function defining it if there is no fear of confusion.

2.1.8 DEFINITION. A C^k function $f(k \geqq 1)$ defined in a neighbourhood W of a is called *stationary* at a if there is a coordinate system (U, φ) with $U \subset W$, $a \in U$ such that all first partial derivatives of $f \circ \varphi^{-1}$ are zero at $\varphi(a)$. A germ of C^k functions is stationary at a if there is an (f, W) in this germ such that f is stationary at a.

Note that if a germ of C^k functions is stationary at a, then any C^k function defining it is stationary at a.

We denote by $C_{a,k}$ the set of all germs of C^k functions at a; $S_{a,k}$ is the set of C^k germs at a which are stationary at a. Let $\boldsymbol{m}_{a,k}$ be the set of C^k germs vanishing at a. $C_{a,k}$ is an R-algebra; $S_{a,k}$, $\boldsymbol{m}_{a,k}$ are subalgebras, and $\boldsymbol{m}_{a,k}$ is even an ideal in $C_{a,k}$. It is the unique maximal ideal of $C_{a,k}$ since any element of $C_{a,k}-\boldsymbol{m}_{a,k}$ is a unit. Furthermore, if $f, g \in \boldsymbol{m}_{a,k}$, then $fg \in S_{a,k}$, and every constant $\in S_{a,k}$.

When the dependence of these spaces on the manifold V is relevant, we denote them by $C_{a,k}(V)$, $S_{a,k}(V)$, $\boldsymbol{m}_{a,k}(V)$ respectively.

We call germs of C^k functions at a also simply C^k germs at a. C^k germs at a can be added, multiplied and composed with C^k maps in the obvious way. Further, the value $g(a)$ of a C^k germ g at a is well defined.

2.1.9 DEFINITION. Let V be a C^k manifold, $k \geqq 1$. The vector space $C_{a,k}/S_{a,k} = T_a^*(V)$ is called the *space of differentials* (or cotangent vectors or covectors) at a. If $f \in C_{a,k}$, its image in $T_a^*(V)$ is denoted by $(df)_a$.

The dual space $T_a(V)$ of $T_a^*(V)$, which can be identified with the set of R linear mappings $X: C_{a,k} \to R$ vanishing on $S_{a,k}$ is called the *tangent space* of V at a. An element of $T_a(V)$ is called a *tangent vector* at a. An R linear function $L: C_{a,k} \to R$ is called a *derivation* if for $f, g \in C_{a,k}$, we have

$$L(fg) = L(f)g(a)+f(a)L(g).$$

2.1.10 PROPOSITION. Any tangent vector $X \in T_a(V)$ is a derivation on $C_{a,k}$.

PROOF. If $f, g \in C_a$, then

$$\varphi = fg - f(a)g - fg(a) \in S_{a,k}$$

[since clearly $\varphi = (f-f(a))(g-g(a))-f(a)g(a)$]. Hence $X(\varphi) = 0$. This means precisely that

$$X(fg) = f(a)X(g) + X(f)g(a).$$

Let $a \in V$ and let (U, φ) be a coordinate system with $a \in U$. If $\varphi = (\varphi_1, \ldots, \varphi_n)$ and $x \in U$, we set $\varphi_j(x) = x_j, j = 1, \ldots, n$. We define, for each j, a tangent vector $(\partial/\partial x_j)_a \in T_a(V)$ by

$$\left(\frac{\partial}{\partial x_j}\right)_a f = \frac{\partial(f \circ \varphi^{-1})}{\partial x_j}(\varphi(a)), \qquad f \in C_{a,k}.$$

(It is obvious that $(\partial/\partial x_j)_a$ are tangent vectors.)

2.1.11 PROPOSITION. $(\partial/\partial x_1)_a, \ldots, (\partial/\partial x_n)_a$ form a basis of $T_a(V)$; in particular, $T_a(V)$ and $T_a^*(V)$ are n-dimensional vector spaces and $T_a^*(V)$ is the dual of $T_a(V)$. Moreover, if $X \in T_a(V), f \in C_{a,k}$, we have $(\mathrm{d}f)_a(X) = X(f)$. [The last remark justifies the notation used.]

PROOF. For $f \in C_{a,k}$, define $g \in C_{a,k}$ by

$$g(x) = f(x) - f(a) - \sum_{j=1}^{n} x_j \left(\frac{\partial}{\partial x_j}\right)_a f.$$

It is clear that $g \in S_{a,k}$; hence, if $X \in T_a(V)$, we have $X(g) = 0$. This gives

$$X(f) = \sum X(x_j) \left(\frac{\partial}{\partial x_j}\right)_a f,$$

i.e.,

$$X = \sum X(x_j) \left(\frac{\partial}{\partial x_j}\right)_a,$$

which means that the $(\partial/\partial x_j)_a$, $1 \leqq j \leqq n$, span $T_a(V)$.

If $X = \sum \lambda_j (\partial/\partial x_j)_a = 0$, we have $\lambda_j = X(x_j) = 0$, since $(\partial/\partial x_j)_a x_k = \delta_{jk}$, (the Kronecker $\delta_{jk} = 0$ if $j \neq k$, $= 1$ if $j = k$). Hence the $(\partial/\partial x_j)_a$ are linearly independent. The last remark is an immediate consequence of the definition.

Dual to the above result, we have the following. If $a \in V$ and

(U, φ) is a coordinate system with $a \in U$, we set, as before, $\varphi(x) = (x_1, \ldots, x_n)$. Then $x_j \in C_{a,k}$, so defines a covector $(dx_j)_a \in T_a^*(V)$.

2.1.11′ PROPOSITION. The $(dx_j)_a$, $1 \leqq j \leqq n$, form a basis of $T_a^*(V)$. This is the basis dual to that given by proposition 2.1.11. Further, if $f \in C_{a,k}$ we have

$$(df)_a = \sum \left\{ \left(\frac{\partial}{\partial x_j} \right)_a f \right\} (dx_j)_a.$$

The proof follows from the obvious fact that $(\partial/\partial x_j)_a$ applied to $(dx_k)_a$, which is $(\partial/\partial x_j)_a x_k$, is equal to δ_{jk}.

2.1.12 REMARK. We note that one can define, also for a *complex-valued* C^k function f its differential $(df)_a$ which is then an element of $T_a^*(V) \otimes_R C$. Further, for any $X \in T_a(V)$, and a complex-valued f, $(df)_a(X)$ is defined as a complex number; moreover if we write $f = f_1 + if_2$ where f_1, f_2 are real valued C^k functions then

$$(df)_a(X) = (df_1)_a(X) + i(df_2)_a(X).$$

It is very useful, for certain purposes, to replace $T_a(V), T_a^*(V)$ by $T_a(V) \otimes_R C$, $T_a^*(V) \otimes_R C$ and operate with complex-valued objects. When it is necessary to do so, we shall denote these spaces by $\mathfrak{T}_a(V)$, $\mathfrak{T}_a^*(V)$ respectively.

Note that we can in the same way apply a vector $X \in T_a(X)$ to a complex valued C^k function f; the map $f \mapsto X(f) = (df)_a(X)$ from the complex valued germs of C^k functions at a to C is then C linear.

2.1.13 REMARK. If V is a C^k manifold $k \geqq 1$, it is, in a natural way, also a C^1 manifold. We can thus consider $C_{a,1} = \{$germs of C^1 functions at $a\}$. We remark that any tangent vector $X \in T_a(V)$ extends to a derivation of $C_{a,1}$ which vanishes on $S_{a,1}$; this is an immediate consequence of proposition 2.1.11. Remark also that if $1 \leqq r \leqq k$, there is a natural inclusion $C_{a,k} \subset C_{a,r}$: $C_{a,k}$ is, of course, a subalgebra. For any $f \in C_{a,r}$, $1 \leqq r \leqq k$, one can again define its image $(df)_a$ in $T_a^*(V)$.

2.1.14 LEMMA. If $f \in S_{a,k}$, we can write

$$f = \sum_{j=1}^{n} g_j h_j + f(a),$$

where g_j, h_j belong to $C_{a,k-1}$ and vanish at a, i.e., $g_j, h_j \in \boldsymbol{m}_{a,k-1}$ $(k \geqq 1)$.

PROOF. It is clear that we may assume that V is a convex neighbourhood of 0 in R^n, and that $a = 0$. Then we have

$$f(x)-f(0) = \int_0^1 \frac{\mathrm{d}}{\mathrm{d}t} f(tx)\mathrm{d}t = \sum_{j=1}^n x_j g_j(x), \qquad g_j(x) = \int_0^1 \frac{\partial f}{\partial x_j}(tx)\mathrm{d}t.$$

Clearly $x_j \in \boldsymbol{m}_{a,k}$ and $g_j \in C_{a,k-1}$. Moreover

$$g_j(0) = \frac{\partial f}{\partial x_j}(0) = 0,$$

since f is stationary by hypothesis.

2.1.15 PROPOSITION. Suppose that X is a derivation of $C_{a,k-1}(k \geqq 1)$. Then its restriction to $C_{a,k}$ is a tangent vector. (X need not be a tangent vector on $C_{a,k-1}$; see PAPY [1956]).

PROOF. We have to prove that $X(f) = 0$ if $f \in S_{a,k}$. Remark first that

$$X(1) = X(1.1) = X(1) \cdot 1 + 1 \cdot X(1),$$

which implies that $X(c) = 0$ for any constant c. It follows now from lemma 2.1.14 that, if $f \in S_{a,k}$ and

$$f - f(a) = \sum_{j=1}^n g_j h_j,$$

where $g_j, h_j \in \boldsymbol{m}_{a,k-1}$, then

$$X(f) = \sum_{j-1}^n (X(g_j)h_j(a) + g_j(a)X(h_j)) = 0.$$

2.1.16 PROPOSITION. If V is a C^∞ manifold and $a \in V$ then $f \in C_{a,\infty}$ is stationary at a if and only if $f - f(a) \in (\boldsymbol{m}_{a,\infty})^2$ [square of the ideal $\boldsymbol{m}_{a,\infty}$]. In particular, $T_a^*(V) = \boldsymbol{m}_{a,\infty}/\boldsymbol{m}_{a,\infty}^2$.

This follows at once from lemma 2.1.14.

Another way of looking at tangent vectors is the following. Let I be the unit interval [0, 1] in R and $\gamma: I \to V$ a C^k curve (i.e., a C^k

map of a neighbourhood of I into V). The tangent to γ at $a = \gamma(0)$ is the element $X = X_\gamma \in T_a(V)$ defined by

$$X(f) = \frac{\mathrm{d}}{\mathrm{d}t} f \circ \gamma(t) \Big|_{t=0}, \qquad f \in C_{a,1}.$$

It is immediate that this defines a tangent vector. Then we have:

2.1.17 Proposition. Any $X \in T_a(V)$ is the tangent at a to a C^k curve γ with $\gamma(0) = a$.

Proof. Let (U, φ) be a coordinate system with $a \in U$, $\varphi(a) = 0$ and

$$\varphi(U) = \{x \in R^n \mid |x_j| < 1\}.$$

Let $X = \sum_{j=1}^n a_j(\partial/\partial x_j)_a$, $a_j \in R$. Let γ_j' be C^k functions in a neighbourhood of I with $|\gamma_j'(t)| < 1$ on I, $\gamma_j'(t) = a_j t$ for t in a neighbourhood of 0. We may take for γ the curve $\varphi^{-1} \circ \gamma'$, where γ' is the map given by $\gamma'(t) = (\gamma_1'(t), \ldots, \gamma_n'(t))$.

§ 2.2 The tangent and cotangent bundles

In this section V will denote a C^k manifold of dimension n with $k \geqq 1$. Let W be another C^k manifold of dimension n and $f\colon V \to W$ a C^k map. Let $b = f(a)$. We define R linear maps:

$$f_{*,a}\colon T_a(V) \to T_a(W), \qquad f_a^*\colon T_b^*(W) \to T_a^*(V)$$

as follows: If $X \in T_a(V)$ and $g \in C_{b,k}(W)$, we set

$$f_{*,a}(X)g = X(g \circ f).$$

[Since $g \circ f \in S_{a,k}(V)$ if $g \in S_{b,k}(W)$, this defines an element of $T_b(W)$]. If $\varphi \in C_{b,k}(W)$ has image $\mathrm{d}\varphi = (\mathrm{d}\varphi)_b \in T_b^*(W)$, we define

$$f_a^*(\mathrm{d}\varphi) = \mathrm{d}(\varphi \circ f)_a.$$

[This map $C_{b,k}(W) \to C_{a,k}(V)$ defines a map $f_a^*\colon T_b^*(W) \to T_a^*(V)$ since $\varphi \circ f \in S_{a,k}(V)$ whenever $\varphi \in S_{b,k}(W)$.] As is easily verified, f_* and f^* are transposes (or adjoints) of each other. f_* is called the differential or tangent map of f at a.

2.2.1 REMARK. If V_1, V_2, V_3 are C^k manifolds and $f_1: V_1 \to V_2$, $f_2: V_2 \to V_3$ are C^k maps, then $f_2 \circ f_1 = f: V_1 \to V_3$ is again a C^k map and we have

$$f_{*,a} = f_{2*,f_1(a)} \circ f_{1*,a}, \qquad f_a^* = f_{1,a}^* \circ f_{2,f_1(a)}^*.$$

It follows that if $f: V \to W$ is a C^k diffeomorphism, then f_a^* is an isomorphism of $T_b^*(W)$ onto $T_a^*(V)$, $b = f(a)$. Apply the formula above to $\mathrm{id}_V = f^{-1} \circ f$, $\mathrm{id}_W = f \circ f^{-1}$. Since, by proposition 2.1.11, $\dim V = \dim_R T_a^*(V)$, we deduce:

2.2.2 COROLLARY. C^k diffeomorphic manifolds ($k \geqq 1$) have the same dimension. (See also remark 2.1.4).

If V is a C^k manifold, let

$$T(V) = \bigcup_{a \in V} T_a(V)$$

be the disjoint union of the $T_a(V)$. There is a natural map $p: T(V) \to V$ which sends $\xi \in T_a(V)$ to the point a; i.e., p is the map for which $p^{-1}(a) = T_a(V)$. We shall prove the following:

2.2.3 THEOREM. $T(V)$ carries a natural structure of a C^{k-1} manifold with respect to which p is a C^{k-1} map. This structure is uniquely determined by the following requirement.

Any $a \in V$ has a neighbourhood U such that there is a C^{k-1} diffeomorphism $h: p^{-1}(U) \to U \times R^n$ such that, if π_1, π_2 denote the projections of $U \times R^n$ onto the first and second factor respectively, we have $\pi_1 \circ h = p$ and $\pi_2 \circ h | T_a(V)$, $a \in V$, is an isomorphism of $T_a(V)$ onto R^n as R vector spaces.

PROOF. Let $\{(U_i, \varphi_i)\}_{i \in \mathscr{I}}$ be a C^k structure on V, and let $\varphi_i(x) = (x_1, \ldots, x_n) \in R^n$; the x_j are C^k functions on U_i. We have seen that any $X \in T_a(V)$ is uniquely determined by the $a_\nu = X(x_\nu)$, $\nu = 1, \ldots, n$ (Proof of proposition 2.1.11), and $X = \sum a_\nu (\partial/\partial x_\nu)_a$. Let now (U_j, φ_j), $j \in \mathscr{I}$, be another coordinate system, such that $a \in U_j$. We set $\varphi_j(x) = (y_1, \ldots, y_n)$ for $x \in U_j$. Now, $a \in U_i \cap U_j$ and $x_1, \ldots, x_n$ are C^k functions on $U_i \cap U_j$. We have, if $g \in C_{a,k}$,

$$\left(\frac{\partial}{\partial y_\mu}\right)_a g = \frac{\partial}{\partial y_\mu}(g \circ \varphi_j^{-1})(\varphi_j(a))$$
$$= \frac{\partial}{\partial y_\mu}(g \circ \varphi_i^{-1} \circ \varphi_i \circ \varphi_j^{-1})(\varphi_j(a))$$
$$= \sum_{\nu=1}^{n} \frac{\partial}{\partial x_\nu} g \circ \varphi_i^{-1}(\varphi_i(a)) \cdot \frac{\partial(\varphi_i \circ \varphi_j^{-1})_\nu}{\partial y_\mu}(\varphi_j(a))$$
$$[\varphi_i \circ \varphi_j^{-1} = ((\varphi_i \circ \varphi_j)_1^{-1}, \ldots)],$$
$$= \sum_{\nu=1}^{n} \left(\frac{\partial}{\partial x_\nu}\right)_a g \cdot b_{\mu\nu}(a)$$

where

$$b_{\mu\nu} = \left(\frac{\partial}{\partial y_\mu}\right)_a x_\nu,$$

since, as we see at once,

$$\left(\frac{\partial}{\partial y_\mu}\right)_a x_\nu = \frac{\partial(\varphi_i \circ \varphi_j^{-1})_\nu}{\partial y_\mu}(\varphi_j(a)).$$

Thus we have

2.2.4 $$\left(\frac{\partial}{\partial y_\mu}\right)_a = \sum_{\nu=1}^{n} b_{\mu\nu}(a) \left(\frac{\partial}{\partial x_\nu}\right)_a, \qquad b_{\mu\nu}(a) = \left(\frac{\partial}{\partial y_\mu}\right)_a x_\nu.$$

Note that $b_{\nu\mu} \in C^{k-1}(U_i \cap U_j)$. If $M_{ji}(a)$ denotes the matrix $(b_{\mu\nu}(a))$, we have $M_{ii}(a) = I$ (unit $n \times n$ matrix) for all $a \in U_i$, $M_{ij}(a)M_{ji}(a) = I$ for $a \in U_i \cap U_j$ and $M_{kj}(a)M_{ji}(a) = M_{ki}(a)$ if $a \in U_i \cap U_j \cap U_k$. Further, if

$$X = \sum a_\nu \left(\frac{\partial}{\partial x_\nu}\right)_a = \sum b_\mu \left(\frac{\partial}{\partial y_\mu}\right)_a,$$

we have

2.2.5 $$(a_1, \ldots, a_n) = (b_1, \ldots, b_n)M_{ji}(a).$$

Consider now the covering $\{p^{-1}(U_i)\}_{i\in\mathscr{I}}$ of $T(V)$. For each i, we define a map

$$h_i\colon p^{-1}(U_i) \to U_i \times R^n,$$

as follows. Let $\varphi_i = (x_1, \ldots, x_n)$ be the coordinates corresponding to U_i, and $X \in T_a(V)$, $a \in U_i$. We set $X = \sum a_\nu(\partial/\partial x_\nu)_a$ and $v =$

$(a_1, \ldots, a_n) \in R^n$. We then set

$$h_i(X) = (a, v).$$

It is trivial that $p = \pi_1 \circ h_i$ on $p^{-1}(U_i)$ and that $\pi_2 \circ h_i | T_a(V)$ is a vector space isomorphism of $T_a(V)$ onto R^n. Hence h_i is a bijection. Moreover, if $U_i \cap U_j \neq 0$, the map

$$h_j \circ h_i^{-1} : (U_i \cap U_j) \times R^n \to (U_i \cap U_j) \times R^n$$

is given by

$$h_j \circ h_i^{-1}(x, v) = (x, vM_{ji}(x)),$$

[by (2.2.5)] and so is a C^{k-1} map, in particular is continuous. Hence, there is a unique topology on $T(V)$ for which the $p^{-1}(U_i)$ are open sets and h_i is a homeomorphism; the map $p: T(V) \to V$ is then continuous. It is further evident (since p is continuous and each $p^{-1}(U_i)$ is Hausdorff) that $T(V)$ is Hausdorff.

Since, in addition the maps $h_j \circ h_i^{-1}$ are C^{k-1}, it is clear that the family $\{p^{-1}(U_i), h_i\}$ makes of $T(V)$ a C^{k-1} manifold. It is obvious, by construction, that the other properties stated in theorem 2.2.3 are satisfied.

Remark: $T(V)$ is an example of a real vector bundle (see § 3.1). In the same way one proves the following result:

2.2.6 THEOREM. The set

$$T^*(V) = \bigcup_{a \in V} T_a^*(V)$$

carries a natural structure of a C^{k-1} manifold.

The dimension of $T(V)$ and $T^*(V)$ is $2n$. Once more, any point of V has a neighbourhood U such that $\bigcup_{a \in U} T_a^*(V)$ is diffeomorphic to a product $U \times R^n$ with properties as in theorem 2.2.3.

2.2.7 DEFINITION. The triple $(T(V), p, V)$, p being the natural projection, is called the *tangent bundle* of V. If $p^*: T^*(V) \to V$ denotes the natural projection (which sends $T_a^*(V)$ onto $\{a\}$), the triple $(T^*(V), p^*, V)$ is called the *cotangent bundle* of V.

We shall often speak of $T(V)$, $T^*(V)$ as the tangent or cotangent bundle respectively, meaning the space as well as the triple.

We note that when V is a real analytic manifold, $T(V)$, $T^*(V)$ are actually again real analytic manifolds and the projection on V as well as the local isomorphisms with $U \times R^n$ can be chosen to be real analytic maps.

Another such 'bundle' which we shall need is obtained as follows. Let $0 \leqq p \leqq n$. We consider the pth exterior power of the vector space $T_a^*(V)$, denoted, as usual, by $\wedge^p T_a^*(V)$. Let (U, φ) be a coordinate system with $a \in U$, $\varphi(x) = (x_1, \ldots, x_n)$. Then $(dx_1)_a, \ldots, (dx_n)_a$ form an R-basis of $T_a^*(V)$ [see proposition 2.1.11′]. Hence a basis of $\wedge^p T_a^*(V)$ is given by the elements

$$(dx_{j_1})_a \wedge \ldots \wedge (dx_{j_p})_a, \qquad j_1 < \ldots < j_p.$$

We set

$$\wedge^p T^*(V) = \bigcup_{a \in V} \wedge^p T_a^*(V).$$

An element of $\wedge^p T_a^*(V)$ is called a p-covector at a. As with theorems 2.2.3 and 2.2.6, we have the following:

2.2.8 THEOREM. $\wedge^p T^*(V)$ carries a natural structure of a C^{k-1} manifold of dimension $n + \binom{n}{p}$.

This is called the *bundle of p-forms on V*. The bundle $\wedge^p \mathfrak{T}^*(V)$ of complex valued p forms on V is defined in the same way.

2.2.9 DEFINITION. Let V and W be C^k manifolds, $k \geqq 1$, and $f: V \to W$ a C^k map. f is said to have *rank r* at $a \in V$ if

$$\operatorname{rank} \{f_{*,a}: T_a(V) \to T_{f(a)}(W)\} = r.$$

We now calculate the map $f_{*,a}$ in terms of local coordinates. Let $b = f(a)$, and let (U, φ) be a coordinate system on V with $a \in U$, (U', φ') one on W with $b \in U'$ such that $f(U) \subset U'$. We write $\varphi(x) = (x_1, \ldots, x_n)$ for $x \in U$ and $\varphi'(y) = (y_1, \ldots, y_m)$ for $y \in U'$. These give rise to bases $(\partial/\partial x_\nu)_a$, $1 \leqq \nu \leqq n$; $(\partial/\partial y_\mu)$, $1 \leqq \mu \leqq m$ of $T_a(V)$ and $T_b(W)$ respectively. Let

$$X \in T_a(V), \qquad X = \sum_{\nu=1}^{n} a_\nu \left(\frac{\partial}{\partial x_\nu}\right)_a$$

and let

$$f_{*,a}(X) = \sum_{\mu=1}^{m} b_\mu \left(\frac{\partial}{\partial y_\mu}\right)_b .$$

Then we have, if $g \in C_{b,k}$

$$\sum_{\mu=1}^{m} b_\mu \left(\frac{\partial}{\partial y_\mu}\right)_b g = X(g \circ f) = \sum_{\nu=1}^{n} a_\nu \sum_{\mu=1}^{m} \left(\frac{\partial}{\partial y_\mu}\right)_b g \cdot \left(\frac{\partial}{\partial x_\nu}\right)_a f_\mu ,$$

where $f_\mu = y_\mu \circ f$. This shows that $f_{*,a}$ is represented by the linear transformation

$$(a_1, \ldots, a_n) \mapsto (b_1, \ldots, b_m),$$

where

$$b_\mu = \sum_{\nu=1}^{n} a_\nu \left(\frac{\partial}{\partial x_\nu}\right)_a f_\mu .$$

This is precisely the map $(\mathrm{d}F)(\varphi(a))$, if we denote by F the map $\varphi' \circ f \circ \varphi^{-1}$ of $\varphi(U)$ into $\varphi'(U')$ (1.3.1).

In view of this remark, we obtain from the results of ch. 1, the following theorems.

2.2.10 Inverse function theorem. If V, W are C^k (real analytic) manifolds of dimension n and f: $V \to W$ is a C^k (real analytic) map and if, for some $a \in V$, $f_{*,a}$: $T_a(V) \to T_{f(a)}(W)$ is an isomorphism, then there exist neighbourhoods U of a and U' of $f(a)$ such that $f|U$ is a C^k (real analytic) isomorphism onto U'.

2.2.11 The rank theorem. Let V, W be C^k (real analytic) manifolds of dimension n and m respectively and suppose that the rank of f at every point $a \in V$ is an integer r independent of a. Then there exist coordinate systems (U, φ), (U', φ') of a and $f(a)$ respectively such that $\varphi' \circ f \circ \varphi^{-1}|\varphi(U)$ is the map

$$(x_1, \ldots, x_n) \mapsto (x_1, \ldots, x_r, 0, \ldots, 0).$$

2.2.12 Definition. Let V be a C^1 manifold of dimension n, which has a countable base. Let $S \subset V$. Then S is said to have *measure zero* if for any coordinate system (U, φ), $\varphi(S \cap U)$ has measure zero in $\mathbf{R}^n$.

We can define critical points as in ch. 1. Let V, W be C^k manifolds of dimension n and m respectively, $k \geqq 1$ and let $f: V \to W$ be a C^k map. A point $a \in V$ is called critical if the rank of f at a is $< m$.

The following theorem follows at once from theorem 1.4.6.

2.2.13 SARD'S THEOREM. If V, W are C^∞ manifolds which possess countable bases and if $f: V \to W$ is a C^∞ map, then the image $f(A)$ of the set A of critical points of f has measure 0 in W.

Finally, as in § 1.2 one can prove the existence of partitions of unity. If (U, φ) is a coordinate system on the C^k manifold V, we deduce at once from lemma 1.2.5 that if K is a compact subset of U, there is a C^k function $\eta \geqq 0$ on V such that $\eta(x) > 0$ for $x \in K$ and $\operatorname{supp}(\eta) \subset U$. Then, exactly as we proved theorem 1.2.3, we prove the following theorem:

2.2.14 THEOREM. Suppose given an open covering $\{V_i\}_{i \in \mathscr{I}}$ of the C^k manifold V, $0 \leqq k \leqq \infty$, and suppose that V has a countable base. Then there exists a family $\{\eta_i\}_{i \in \mathscr{I}}$ of C^k functions, $\eta_i \geqq 0$, $\operatorname{supp}(\eta_i) \subset V_i$ such that the family $\{\operatorname{supp}(\eta_i)\}$ is locally finite and $\sum \eta_i(x) = 1$ for any $x \in V$.

2.2.15 COROLLARY. If X is a closed set in V and $U \supset X$ is open, there exists a C^k function η on V with $\eta(x) = 1$ if $x \in X$ and $\eta(x) = 0$ if $x \in V - U$.

§ 2.3 Grassmann manifolds

In this section we give a class of examples of real (and complex) analytic manifolds which are of importance in many contexts, although they will not be used in the rest of this book.

Let E be a vector space of dimension n over R. Let r be an integer, $0 < r < n$. We denote by $G_r(E)$ the set of r-dimensional linear subspaces of E. We shall show that $G_r(E)$ carries a natural structure of a real analytic manifold.

Let F be any subspace of dimension r. We can then find a subspace E' of E of dimension $n-r$ such that $F \cap E' = \{0\}$. We then have $E = F + E'$. It follows that if, for a subspace E' of E of dimension $n-r$, we set

$$U(E') = \{F|F \text{ of dimension } r,\ F \subset E, F \cap E' = \{0\}\},$$

then

$$G_r(E) = \bigcup_{E'} U(E').$$

We claim now that $U(E')$ may be identified with the set $L(E')$ of all R linear maps $\lambda: E/E' \to E$ such that $\eta_{E'} \circ \lambda =$ identity, where $\eta_{E'}: E \to E/E'$ is the natural projection. In fact $[\lambda \in L(E')]$ we clearly have: $\lambda(E/E') = F$ is a subspace of E of dimension r such that $F \cap E' = \{0\}$. Conversely, if F is given, then

$$E = F + E'.$$

Denote by π_F the projection on F given by the decomposition $E = F + E'$. Then

$$\pi_F(E') = 0,$$

hence induces an R linear map

$$\lambda'_F: E/E' \to F,$$

hence a map

$$\lambda_F: E/E' \to E.$$

Clearly

$$\eta_{E'} \circ \lambda_F = \text{identity}$$

and

$$\lambda_F(E/E') = F.$$

This shows that $U(E')$ and $L(E')$ can be identified naturally.

We next claim that $L(E')$ is a finite-dimensional affine space, i.e., for any $\lambda_0 \in L(E')$, the set $\{\lambda - \lambda_0 | \lambda \in L(E')\}$ is a finite-dimensional vector space. In fact, if λ_0, $\lambda \in L(E')$, then $\eta_E \circ (\lambda - \lambda_0) = 0$, hence

$$(\lambda - \lambda_0)(E/E') \subset \ker(\eta_E) = E'$$

and $\lambda - \lambda_0$ is an R linear map of E/E' into E', i.e.,

$$\lambda - \lambda_0 \in \mathrm{Hom}_R(E/E', E').$$

Further, if $\mu \in \mathrm{Hom}_R(E/E', E')$, then $\lambda_0 + \mu \in L(E')$. This shows that $\{\lambda - \lambda_0 | \lambda \in L(E')\}$ is naturally isomorphic with $\mathrm{Hom}_R(E/E', E')$, which proves our result. Now, if we choose a basis for E' and complete it to one for E, we can identify $\mathrm{Hom}_R\,(E/E', E')$ with the set $\mathrm{Hom}_R(R^r, R^{n-r}) = M(r, n-r)$, the set of $r\times(n-r)$ matrices, which is $\simeq R^{r(n-r)}$.

Thus, *given* $\lambda_0 \in U(E')$ *and suitable bases for* E' *and* E, *there is a bijection* $h_{E'}$ *of* $U(E')$ *onto* $R^{r(n-r)}$.

Further, it can be shown quite easily that for two subspaces E', E'' of dimension $n-r$, the map $h_{E''} \circ h_{E'}^{-1}$ of $h_{E'}(U(E') \cap U(E''))$ into $R^{r(n-r)}$ is given by *rational functions* (in particular, real analytic functions). It follows that there is a unique topology on $G_r(E)$ with respect to which the $U(E')$ are open sets and the mappings $h_{E'}$ are homeomorphisms.

We claim that this topology is Hausdorff. First, $U(E')$ is Hausdorff and open for any E'. Moreover, given $F_1, F_2 \subset E$ which are subspaces of dimension r, we can find a subspace E' of dimension $n-r$ so that $F_1 \cap E'$, $F_2 \cap E'$ both reduce to $\{0\}$, so that $F_1, F_2 \in U(E')$. It follows at once that if $F_1 \neq F_2$, they have disjoint neighbourhoods in $U(E')$, and hence in $G_r(E)$.

With these remarks, we have shown that $G_r(E)$ carries a structure of a real analytic manifold. We next show that it is compact.

Let $e_1, \ldots, e_n$ be a basis of E_1 and let $O(n)$ be the orthogonal group of R^n, i.e., the set of $n\times n$ real matrices such that

$$A \cdot {}^tA = I$$

where ${}^tA$ denotes the transpose of the matrix A and I is the unit $n\times n$ matrix. We have an isomorphism $\varphi\colon E \to R^n$ given by

$$\varphi(x) = (x_1, \ldots, x_n) \qquad \text{if} \quad x = \sum_{j=1}^{n} x_j e_j.$$

Hence $O(n)$ acts on E; if $A \in O(n)$ and $x \in E$, we set

$$A(x) = \varphi^{-1} \cdot A \cdot \varphi(x),$$

where $A \cdot \varphi(x)$ is the usual action of $O(n)$ on R^n. Hence it acts also on $G_r(E)$. We claim that given $F_1, F_2 \in G_r(E)$, there is an $A \in O(n)$ with $A \cdot F_1 = F_2$. We have only to prove it for $E = R^n$, where it is obvious [since we can choose orthonormal bases $(u_1, \ldots, u_n)$, $(v_1, \ldots, v_n)$ of R^n such that F_1 is the space spanned by $(u_1, \ldots, u_r)$, F_2 that spanned by $(v_1, \ldots, v_r)$]. Further the map $O(n)\times G_r(E)$

$\to G_r(E)$ given by $(A, F) \to A \cdot F$ is continuous. Moreover $O(n)$ is compact. If $F_0 \in G_r(E)$ is fixed, $G_r(E)$ is the image of $O(n)$ under the continuous map $A \mapsto A \cdot F_0$, and so is compact. Thus we have proved:

2.3.1 THEOREM. The Grassmann manifold $G_r(E)$ is a compact real analytic manifold of dimension $r(n-r)$.

2.3.2 REMARK. If $r = 1$, $G_1(E)$ is called the projective space $P(E)$ associated with E; $P(E)$ has dimension $n-1$. If, in particular $E = R^n$, we write RP^{n-1} for $P(R^n)$.

2.3.3 REMARK. If E is an n-dimensional complex vector space, $0 < r < n$, $\mathscr{G}_r(E)$ denotes the set of *complex* r-dimensional subspaces of E. In the same way as above, we prove that $\mathscr{G}_r(E)$ is a compact complex manifold of complex dimension $r(n-r)$. Again when $r = 1$, this is called the projective space $P(E)$ of E, and if $E = C^n$, we write CP^{n-1} for $P(E)$.

§ 2.4 Vector fields and differential forms

Let V be a C^k manifold and $T(V)$ its tangent bundle. There is a natural projection $p: T(V) \to V$ which maps $T_a(V)$ onto the point a.

2.4.1 DEFINITION. A *vector field* on V is a map $X: V \to T(V)$ such that $p \circ X =$ identity on V. If X is a C^r map $(0 \leqq r \leqq k-1)$, we call X a C^r vector field. If V is an analytic manifold, analytic vector fields are similarly defined.
In general, if $p: B \to A$ is a map of the set B onto the set A, a map $X: A \to B$ such that $p \circ X =$ identity of A is called a cross section of p. Vector fields are thus cross sections of the projection of the tangent bundle onto the manifold.

If X is a vector field and (U, φ) is a coordinate system, $\varphi(x) = (x_1, \ldots, x_n)$, then we can write

$$X(a) = \sum_{\nu=1}^{n} \xi_\nu(a) \left(\frac{\partial}{\partial x_\nu}\right)_a, \qquad \xi_\nu(a) \in R.$$

It is easily verified that X is a C^r vector field if and only if the ξ_ν are C^r functions on U for any coordinate system (U, φ). We write $\partial/\partial x_\nu$ for the vector field $a \leftrightarrow (\partial/\partial x_\nu)_a$. Let p be an integer between 0 and n, $0 \leqq p \leqq n$ and $\wedge^p T^*(V)$ the bundle of p-forms on V.

2.4.2 DEFINITION. A *differential form of degree p* (or simply, a p-form) is a map ω: $V \to \wedge^p T^*(V)$ such that for any $a \in V$ we have

$$\omega(a) \in \wedge^p T_a^*(V).$$

A p-form is of class C^r(analytic) if the map ω has this property. Thus, differential forms are cross sections of the natural projection of $\wedge^p T^*(V)$ onto V.

If (U, φ) is a coordinate system, and $\varphi(x) = (x_1, \ldots, x_n)$ then the elements $(\mathrm{d}x_1)_a, \ldots, (\mathrm{d}x_n)_a$ form a basis of $T_a^*(V)$ and a basis of $\wedge^p T_a^*(V)$ is given by the elements $(\mathrm{d}x_{j_1})_a \wedge \ldots \wedge (\mathrm{d}x_{j_p})_a$ $1 \leqq j_1 < \ldots < j_p \leqq n$. Hence, a p-form ω has a unique representation

$$\omega(a) = \sum_{j_1 < \ldots < j_p} \xi_{j_1 \ldots j_p}(a)(\mathrm{d}x_{j_1})_a \wedge \ldots \wedge (\mathrm{d}x_{j_p})_a.$$

Again, ω is of class C^r (analytic) if and only if the $\xi_{j_1} \ldots_{j_p}$ are, for all coordinate systems (U, φ). We write $\mathrm{d}x_j$ for the 1-form

$$a \leftrightarrow (\mathrm{d}x_j)_a,$$

$\mathrm{d}x_{j_1} \wedge \ldots \wedge \mathrm{d}x_{j_p}$ for the p-form

$$a \leftrightarrow (\mathrm{d}x_{j_1})_a \wedge \ldots \wedge (\mathrm{d}x_{j_p})_a.$$

Note that for integers p, $q \geqq 0$, we have a natural linear map

$$\wedge^p T_a^*(V) \otimes \wedge^q T_a^*(V) \to \wedge^{p+q} T_a^*(V);$$

we denote the image of $\omega_1(a) \otimes \omega_2(a)$ under this map by $\omega_1(a) \wedge \omega_2(a)$. Given now two differential forms ω_1, ω_2 of degree p, q respectively, we can define a $(p+q)$-form $\omega_1 \wedge \omega_2$, called the exterior product of ω_1 and ω_2 by

$$(\omega_1 \wedge \omega_2)(a) = \omega_1(a) \wedge \omega_2(a).$$

It is immediately clear that if ω_1, ω_2 are C^r(analytic) so is $\omega_1 \wedge \omega_2$.

We denote the set of all p-forms on V (of class C^{k-1}) by $A^p(V)$. We will also consider the direct sum

$$A(V) = \sum_{p \geqq 0} A^p(V).$$

The exterior product defined above transforms this into an R-algebra. Elements of $A(V)$ will be called simply *differential forms.* They have components which are p-forms, all but finitely many of these components being zero.

Our formulae in local coordinates then imply that if V is C^k diffeomorphic to an open set in R^n, then $A(V)$ is generated, as an algebra, by elements of the form f, $dg, f \in C^{k-1}(V)$ and $g \in C^k(V)$.

We note that if R is the ring of C^{k-1} functions on V, then the R vector space $\mathfrak{X} = \mathfrak{X}(V)$ of C^{k-1} vector fields on V is also an R-module. Now there is a pairing between $\wedge^p T_a^*(V)$ and $\wedge^p T_a(V)$ which makes them dual spaces to one another. Hence, if ω is a C^{k-1} p-form on V, it defines a multilinear map of $\mathfrak{X}^p$ into R by the formula

$$\omega(X_1, \ldots, X_p)(a) = \omega(a)(X_1(a), \ldots, X_p(a)).$$

This map is alternate and R multilinear (as is evident).

2.4.3 PROPOSITION. Conversely, any alternate, R multilinear map φ of $\mathfrak{X}^p$ into R defines a C^{k-1} p-form ω.

PROOF. To show this, let $X_{1,a}, \ldots, X_{p,a} \in T_a(V)$, $a \in V$. We assert first that there are C^{k-1} vector fields $X_1, \ldots, X_p$ with $X_j(a) = X_{j,a}$. To prove this, suppose $X_{j,a} = \sum a_\nu(\partial/\partial x_\nu)_a$ with respect to a coordinate system (U, φ) and let η be a C^k function with compact supp $(\eta) \subset U$ and $\eta(x) = 1$ for x in a neighbourhood of a (corollary 2.2.15). We may define $X_j \colon V \to T(V)$ by

$$X_j(x) = 0 \qquad x \notin U,$$

$$X_j(x) = \sum \eta(x) \cdot a_\nu \left(\frac{\partial}{\partial x_\nu}\right)_x \qquad x \in U.$$

We next claim that if $X_j(a) = 0$ for some j, we have $\varphi(X_1, \ldots, X_p)(a) = 0$. To see this, since φ is R multilinear, we have only to show that if $X_j(a) = 0$, then $X_j = \sum_{\mu=1}^N \eta_\mu Y_\mu$, where the $\eta_\mu \in C^{k-1}(V)$ and vanish at a and $Y_\mu \in \mathfrak{X}$. Let (U, φ) be a coordinate system with $a \in U$

and let

$$X_j(x) = \sum_{\nu=1}^{n} a_\nu(x) \left(\frac{\partial}{\partial x_\nu}\right)_x, \qquad x \in U.$$

Since $X_j(a) = 0$, we have $a_\nu(a) = 0$, $1 \leqq \nu \leqq n$. Let $\eta \in C^{k-1}(V)$ be such that $\eta(a) = 1$, $0 \leqq \eta \leqq 1$ and supp (η) is a compact subset of U (see corollary 2.2.15). Define vector fields Z_ν, $1 \leqq \nu \leqq n$ by

$$Z_\nu(x) = 0 \quad \text{if} \quad x \notin U, \qquad Z_\nu(x) = b_\nu(x) \left(\frac{\partial}{\partial x_\nu}\right)_x \quad \text{if} \quad x \in U,$$

where

$$b_\nu(x) = \begin{cases} \eta(x) \cdot a_\nu(x), & x \in U; \\ 0 & x \notin U, \end{cases}$$

so that $b_\nu \in C^{k-1}(V)$ and $b_\nu(a) = 0$. We define vector fields Y_ν, by

$$Y_\nu(x) = 0 \quad \text{if} \quad x \notin U, \qquad Y_\nu(x) = \eta(x) \left(\frac{\partial}{\partial x_\nu}\right)_x \quad \text{if} \quad x \in U.$$

We then have

$$X_j = \sum_{\nu=1}^{n} \{b_\nu Y_\nu + (1-\eta) Z_\nu\} + (1-\eta) X_j.$$

Since $\eta(a) = 1$ and $b_\nu(a) = 0$, this has the required form. Now define $\omega(a)$, for any $a \in V$, by

$$\omega(a)(X_{1,a}, \ldots, X_{p,a}) = \varphi(X_1, \ldots, X_p)(a).$$

In view of our remark above, this is zero if any X_j vanishes at a, so that $\omega(a)$ is defined independently of the choice of the X_j with $X_j(a) = X_{j,a}$. An easy computation with local coordinates shows that the element $\omega(a) \in \wedge^p T_a^*(V)$ so defined defines a C^{k-1} p-form ω. Note that a (C^k, analytic) 0-form is simply a (C^k, analytic) function on V.

Let now V, W be two C^k manifolds and f: $V \to W$ a C^k map. Let $a \in V$ and $f(a) = b$. We have defined a map

$$f_a^*: T_b^*(W) \to T_a^*(V).$$

If $p > 0$, this 'extends' to a linear map

$$f_a^*: \wedge^p T_b^*(W) \to \wedge^p T_a^*(V).$$

This gives us an algebra homomorphism, again denoted f_a^*,

$$f_a^*: \wedge T_b^*(W) \to \wedge T_a^*(V).$$

2.4.4 DEFINITION. If ω is a C^{k-1} p-form on W, we define the *pull-back* (or inverse image) $f^*(\omega)$, a C^{k-1} p-form on V, as follows: If $p = 0$, $\omega = g$ is a C^{k-1} function, and we set $f^*(\omega) = g \circ f$. If $p > 0$, we set

$$f^*(\omega)(a) = f_a^*(\omega(f(a))).$$

Note that the induced map

$$f^*: A(W) \to A(V)$$

on the space of all differential forms is an *algebra homomorphism*.

If U is an open set in the C^k manifold V and $i: U \to V$ is the injection then, if ω is a p-form on V, we write $i^*(\omega) = \omega|U$ and call it the restriction of ω to V.

2.4.5 REMARK. We have also the map, for any $a \in V, f_{*,a}: T_a(V) \to T_b(W)$, $b = f(a)$. This however does not in general 'extend' to a map $\mathfrak{X}(V) \to \mathfrak{X}(W)$; for example if $X \in \mathfrak{X}(V)$ and a, a' are points in X with $f(a) = f(a') = b$, but $f_{*,a}(X(a)) \neq f_{*,a'}(X(a'))$. If, however $f: V \to W$ is a C^k diffeomorphism, it is clear that the map

$$b \mapsto f_{*,a}(X(a)), \qquad a = f^{-1}(b)$$

is a C^{k-1} vector field $f_*(X)$ on W.

Let now V be a C^k manifold with $k \geqq 2$. Let X and Y be two C^{k-1} vector fields on V. For any $f \in C^k(V)$, $X(f)$ is a function in $C^{k-1}(V)$. Since $k-1 \geqq 1$, we can apply Y to $X(f)$ (see remark 2.1.13). Thus we can define a map $C^k(V) \to C^{k-2}(V)$ by

$$[X, Y](f) = X(Y(f)) - Y(X(f)).$$

In terms of local coordinates, if (U, φ) is a coordinate system, $\varphi(x) = (x_1, \ldots, x_n)$ and we have

$$X(x) = \sum a_\nu(x) \left(\frac{\partial}{\partial x_\nu}\right)_x, \qquad Y(x) = \sum b_\nu(x) \left(\frac{\partial}{\partial x_\nu}\right)_x,$$

where

$$a_\nu, b_\nu \in C^{k-1}(U);$$

it is easily verified that

$$[X, Y](f)(x) = \sum c_\nu(x) \left(\frac{\partial}{\partial x_\nu}\right)_x,$$

where

$$c_\nu(x) = \sum_\mu \left\{ a_\mu(x) \frac{\partial b_\nu}{\partial x_\mu} - b_\mu(x) \frac{\partial a_\nu}{\partial x_\mu} \right\}.$$

This shows that $[X, Y]$ is once more a vector field, of class C^{k-2}.

The vector field $[X, Y]$ is called the *Poisson bracket* (or just *bracket*) of the vector fields X, Y.

Note that $[X, Y](a)$ depends on the values of X and Y in a whole neighbourhood of a and not just on $X(a)$ and $Y(a)$. We have, obviously, $[X, X] = 0$, $[X, Y] = -[Y, X]$. It is easily checked that if $k \geqq 3$, and X, Y, Z are C^{k-1} vector fields, we have

$$[X, [Y, Z]] + [Y, [Z, X]] + [Z, [X, Y]] = 0.$$

This is called the *Jacobi identity*. Note that if X_ν is the vector field in a coordinate neighbourhood U defined by

$$X_\nu(x) = \left(\frac{\partial}{\partial x_\nu}\right)_x, \qquad 1 \leqq \nu \leqq n,$$

we have $[X_\nu, X_\mu] = 0$ in U for any pair of μ, ν.

2.4.6 We now consider the tangent space and so on of complex manifolds. Let V be a complex manifold of complex dimension n, and let $a \in V$. Let (U, φ) be a coordinate system with $a \in U$; φ is then a complex analytic isomorphism of U with an open set in C^n and we write $\varphi(z) = (z_1, \ldots, z_n)$ for $z \in U$, $z_j = x_j + iy_j$ where x_j, y_j are C^∞ real valued functions. Then the tangent space $T_a(V)$, V being considered as a C^∞ real manifold of dimension $2n$, has a natural structure of a complex vector space which is obtained as follows:
We consider $C^n \simeq R^{2n}$ by the isomorphism

$$(z_1, \ldots, z_n) \leftrightarrow (x_1, y_1, \ldots, x_n, y_n),$$

where $z_j = x_j + iy_j$. Then the mapping $T_a(V) \to R^{2n}$ defined by

$$X \leftrightarrow ((dx_1)_a(X), (dy_1)_a(X), \ldots, (dx_n)_a(X), (dy_n)_a(X)),$$

is an R-isomorphism. Hence the map

$$X \leftrightarrow ((dz_1)_a(X). , \ldots, (dz_n)_a(X))$$

is an R-isomorphism of $T_a(V)$ onto C^n which makes of $T_a(V)$ a C vector space (remark 2.1.12). This complex structure on $T_a(V)$ is independent of the coordinate system (U, φ) chosen. It is, in fact, characterized by the following property. If f is a germ of holomorphic functions at $a \in V$, we have

$$(\mathrm{d}f)_a(\zeta X) = \zeta \cdot (\mathrm{d}f)_a(X), \qquad \zeta \in C, \quad X \in T_a(V),$$

where ζX stands for the operation given by the complex structure on $T_a(V)$ and the term on the right is the product of two complex numbers.

If $\varphi(z) = (z_1, \ldots, z_n)$ are complex coordinates in U and $z_j = x_j + iy_j$ as before, then $(x_1, y_1, \ldots, x_n, y_n)$ form coordinates in U for V considered as a real C^∞ manifold. Thus, we have vectors

$$\left(\frac{\partial}{\partial x_\nu}\right)_a, \left(\frac{\partial}{\partial y_\nu}\right)_a \in T_a(V).$$

One verifies easily that, with respect to the above complex structure on $T_a(V)$, we have

$$i\left(\frac{\partial}{\partial y_\nu}\right)_a = -\left(\frac{\partial}{\partial x_\nu}\right)_a.$$

Moreover, $(\partial/\partial x_\nu)_a z_\mu = \delta_{\mu\nu}$, so that the $(\partial/\partial x_\nu)_a$, $1 \leqq \nu \leqq n$ form a C basis of $T_a(V)$.

Let now (U', φ') be another coordinate system with $a \in U'$ and $\varphi'(z) = (w_1, \ldots, w_n)$, $z \in U$. We write $w_j = u_j + iv_j$ where u_j, v_j are real C^∞ functions.

Let $X \in T_a(V)$ and

$$X = \sum_{\nu=1}^{n} a_\nu \left(\frac{\partial}{\partial x_\nu}\right)_a = \sum_{\mu=1}^{n} b_\mu \left(\frac{\partial}{\partial u_\mu}\right)_a, \qquad a_\nu, b_\mu \in T_a(V).$$

One verifies easily that we have

$$b_\mu = \sum_{\nu=1}^{n} a_\nu \left(\frac{\partial}{\partial x_\nu}\right)_a w_\mu .$$

Further, the functions $(\partial/\partial x_\nu)_a w_\mu$ are *holomorphic functions of* a for $a \in U \cap U'$. We can therefore repeat the proof of theorem 2.2.3 to prove the following:

2.4.7 THEOREM. If V is a complex manifold of dimension n, $T(V)$

$= \bigcup T_a(V)$ carries a natural structure of a complex manifold of dimension $2n$. The projection $p: T(V) \to V$ is holomorphic and $T(V)$ is locally trivial, i.e., any $a \in V$ has a neighbourhood U, such that there is a complex analytic isomorphism $h: p^{-1}(U) \to U \times C^n$ and if $h(\xi) = (a, v)$, where $\xi \in p^{-1}(U)$, one has $p(\xi) = a$. Further, the map $T_a(V) \to C^n$ given by $\xi \mapsto v$ is a C-isomorphism.

2.4.8 REMARK. If V, W are complex manifolds and $f: V \to W$ is a holomorphic map, the map $f_a^*: T_a(V) \to T_{f(a)}(W)$ is a C linear map.

We note that in view of the above remarks, we can apply the rank theorem for holomorphic functions (see remark 1.3.15) to obtain:

2.4.9 RANK THEOREM FOR HOLOMORPHIC MAPPINGS. Let V, W be complex manifolds of dimension n, m respectively and $f: V \to W$ a holomorphic map. Let r be an integer such that, for every $a \in V$,

$$\operatorname{rank}_C f_{a*} = r.$$

Then there exist complex coordinates (U, φ) at a, (U', φ') at $f(a)$ such that $\varphi' \circ f \circ \varphi^{-1} | \varphi(U)$ is the map $(z_1, \ldots, z_n) \mapsto (z_1, \ldots, z_r, 0, \ldots, 0)$.

2.4.10 REMARKS. We make a few remarks on *complex* valued C^∞ differential forms on a complex manifold V. We have identified C^n with R^{2n} by the map

$$(z_1, \ldots, z_n) \mapsto (x_1, y_1, \ldots, x_n, y_n),$$

where $z_j = x_j + iy_j$. Let E be a vector space over C, of dimension n. Consider the C vector space

$$\mathscr{E}^* = \operatorname{Hom}_R(E, C)$$

of R linear maps of E into C. Let

$$F = \{f \in \mathscr{E}^* |\ f(iv) = if(v) \quad \text{for all} \quad v \in E\},$$

$$\bar{F} = \{f \in \mathscr{E}^* |\ f(iv) = -if(v) \quad \text{for all} \quad v \in E\}.$$

Clearly $F, \bar{F}$ are C subspaces of $\mathscr{E}^*$ and we have

$$\mathscr{E}^* = F \oplus \bar{F}.$$

In fact, $F \cap \bar{F} = \{0\}$ and if $g \in \mathscr{E}^*$, we set

$$f'(v) = \tfrac{1}{2}(g(v) - \mathrm{i}g(\mathrm{i}v)), \qquad f''(v) = \tfrac{1}{2}(g(v) + \mathrm{i}g(\mathrm{i}v)).$$

Then $f' \in F$, $f'' \in \bar{F}$ and $g = f' + f''$. We set $F = \mathscr{E}^*_{1,0}$ and $\bar{F} = \mathscr{E}^*_{0,1}$. The map $z \mapsto \bar{z}$ of C onto itself induces an R linear map of $\mathscr{E}^*$ onto itself which takes F to $\bar{F}$. We denote the image of $g \in \mathscr{E}^*$ under this map again by $\bar{g}$. If $(e_1, \ldots, e_n)$ is a C basis of F, $(\bar{e}_1, \ldots, \bar{e}_n)$ is a C basis of $\bar{F}$.

Consider now the exterior product $\wedge^r \mathscr{E}^*$ of $\mathscr{E}^*$ considered as a complex vector space. Let p, q be integers with $p+q = r$, and let $\mathscr{E}^*_{p,q}$ denote the subspace of $\wedge^r \mathscr{E}^*$ generated by elements of the form

$$u_{j_1} \wedge \ldots \wedge u_{j_p} \wedge \bar{v}_{k_1} \wedge \ldots \wedge \bar{v}_{k_q},$$

where $u_{j_\nu} \in F$, $\nu = 1, \ldots, p$, $\bar{v}_{k_\mu} \in \bar{F}$. A basis of $\mathscr{E}^*_{p,q}$ is given by the elements

$$e_{j_1} \wedge \ldots \wedge e_{j_p} \wedge \bar{e}_{k_1} \wedge \ldots \wedge \bar{e}_{k_q} \equiv e_J \wedge \bar{e}_K$$

where $j_1 < \ldots < j_p$, $k_1 < \ldots < k_q$ (but there is no relation between the j and the k).

We have

$$\wedge^r \mathscr{E}^* = \sum_{p+q=r} \mathscr{E}^*_{p,q},$$

the sum being direct.

If V is a complex manifold of dimension n, and $a \in V$, we take $E = T_a(V)$ with the complex structure of $T_a(V)$ defined earlier. Let

$$\mathfrak{T}^*_a(V) = \mathrm{Hom}_R(T_a(V), C) = T^*_a(V) \otimes_R C;$$

this is a C vector space of dimension $2n$. As before, if $(z_1, \ldots, z_n)$ are coordinates in an open set U, $z_j = x_j + \mathrm{i}y_j$, we have

$$(\mathrm{d}x_j)_a, (\mathrm{d}y_j)_a \in T^*_a(V) \subset \mathfrak{T}^*_a(V),$$

and we can define

$$(\mathrm{d}z_j)_a = (\mathrm{d}x_j)_a + \mathrm{i}(\mathrm{d}y_j)_a \in \mathfrak{T}^*_a(V)$$

(see also remark 2.1.12); we also set

$$(\mathrm{d}\bar{z}_j)_a = (\mathrm{d}x_j)_a - \mathrm{i}(\mathrm{d}y_j)_a.$$

One verifies easily that the $(dz_j)_a$, $(d\bar{z}_j)_a$ form a C basis of $\mathfrak{T}_a^*(V)$ (since the $(dx_j)_a$, $(dy_j)_a$ form an R basis of $T_a^*(V)$).

Define

$$\mathfrak{T}_a(V) = T_a(V) \otimes_R C = \mathrm{Hom}_C(\mathfrak{T}_a^*, C).$$

This being the dual of $\mathfrak{T}_a^*(V)$, there is a basis of $\mathfrak{T}_a(V)$ dual to the one given above. If we set

$$\left(\frac{\partial}{\partial z_\nu}\right)_a = \frac{1}{2}\left\{\left(\frac{\partial}{\partial x_\nu}\right)_a - \mathrm{i}\left(\frac{\partial}{\partial y_\nu}\right)_a\right\},$$

$$\left(\frac{\partial}{\partial \bar{z}_\nu}\right)_a = \frac{1}{2}\left\{\left(\frac{\partial}{\partial x_\nu}\right)_a + \mathrm{i}\left(\frac{\partial}{\partial y_\nu}\right)_a\right\},$$

(multiplication by i being in $T_a(V) \otimes_R C$ and *not* in $T_a(V)$ i.e., $\mathrm{i}(\partial/\partial y_\nu)_a = (\partial/\partial y_\nu)_a \otimes i$). Then this dual basis is given precisely by the vectors $(\partial/\partial z_\nu)_a$, $(\partial/\partial \bar{z}_\nu)_a (1 \leqq \nu \leqq n)$. One sees, as before that

$$\mathfrak{T}(V) = \bigcup_{a \in V} \mathfrak{T}_a(V), \qquad \mathfrak{T}^*(V) = \bigcup_{a \in V} \mathfrak{T}_a^*(V)$$

are C^∞ manifolds of real dimension $6n$. $\wedge^p \mathfrak{T}(V)$, $\wedge^p \mathfrak{T}^*(V)$ are defined as before, and a C^∞ complex valued differential p-form ω on V is a C^∞ map ω: $V \to \wedge^p \mathfrak{T}^*(V)$ with $\omega(a) \in {}^p\mathfrak{T}_a^*(V)$ for $a \in V$.

If $E = T_a(V)$, we consider

$$\wedge^r \mathscr{E}^* = \sum_{p+q=r} \mathscr{E}^*_{p,q}$$

as above. We write $\mathscr{E}^*(V, a)$, $\mathscr{E}^*_{p,q}(V, a)$, . . . when the dependence on V and a is relevant. One sees at once that $\mathscr{E}^*_{1,0}$ is spanned by $(dz_1)_a$, . . ., $(dz_n)_a$, and $\mathscr{E}^*_{0,1}$ by $(d\bar{z}_1)_a$, . . ., $(d\bar{z}_n)_a$. Hence $\mathscr{E}^*_{p,q}$ is spanned by the covectors

$$dz_J \wedge d\bar{z}_K = dz_{j_1} \wedge \ldots \wedge dz_{j_p} \wedge d\bar{z}_{k_1} \wedge \ldots \wedge d\bar{z}_{k_q}.$$

Elements of $\mathscr{E}^*_{p,q}$ are called covectors of type (p, q).

2.4.11 DEFINITION. A complex valued differential form ω: $V \to \wedge^r \mathfrak{T}^*(V)$ is said to be of type (p, q) if, for any $a \in V$, $\omega(a) \in \mathscr{E}^*_{p,q}(V, a)$.

In terms of local coordinates, ω is of type (p, q) if and only if

$$\omega(a) = \sum_{\substack{j_1 < \ldots < j_p,\\ k_1 < \ldots < k_q}} \omega_{JK} \mathrm{d}z_J \wedge \mathrm{d}\bar{z}_K, \quad J = (j_1, \ldots, j_p), \quad K = (k_1, \ldots, k_q).$$

If V, W are complex manifolds and $f\colon V \to W$ a holomorphic map, one has again the map

$$f_a^*\colon \mathfrak{T}_b^*(W) \to \mathfrak{T}_a^*(V), \qquad b = f(a)$$

and we can define the pull-back $f^*(\omega)$ of a complex-valued p-form on W as in definition 2.4.4.

2.4.12 Proposition. If f is a holomorphic map, if $a \in V$ and $f(a) = b$, we have

$$f_a^*(\mathscr{E}_{p,q}^*(W, b)) \subset \mathscr{E}_{p,q}^*(V, a).$$

In particular, $f^*(\omega)$ is of type (p, q) if ω is a form of type (p, q) on W.

Proof. We have only to prove that

$$f_a^*(\mathscr{E}_{1,0}^*(W, b)) \subset \mathscr{E}_{1,0}^*(V, a).$$

But this is obvious: if g is a germ of holomorphic functions at b, then

$$f_a^*((\mathrm{d}g)_b) = \mathrm{d}(g \circ f)_a \in \mathscr{E}_{1,0}^*(V, a)$$

since $g \circ f$ is holomorphic.

2.4.13 Remark. We can define *holomorphic vector fields* X on a complex manifold V as holomorphic maps $X\colon V \to T(V)$, (see theorem 2.4.7) such that $X(a) \in T_a(V)$ for any $a \in V$. Now, for any $a \in V$, $T_a(V)$ is, in a natural way, a subspace of $\mathfrak{T}_a(V)$. A vector field $X\colon V \to \mathfrak{T}(V)$ is holomorphic (in the sense that $X(a) \in T_a(V)$ for any a and X represents a holomorphic map into $T(V)$), if and only if to every point $a \in V$ there is a coordinate system (U, φ), $a \in U$ such that in the expression

$$X(b) = \sum_{\nu=1}^{n} \left\{ a_\nu(b) \left(\frac{\partial}{\partial z_\nu}\right)_b + a'_\nu(b) \left(\frac{\partial}{\partial \bar{z}_\nu}\right)_b \right\}, \qquad b \in U,$$

all the $a'_\nu(b)$ are zero and the $a_\nu(b)$ are holomorphic functions of b.

Note, moreover, that if X, Y are holomorphic vector fields, so is $[X, Y]$.

§ 2.5 Submanifolds

Let V be a C^k manifold with $k \geqq 1$. Consider the set of C^k manifolds W and injective (i.e., one-one) C^k mappings i: $W \to V$ such that, for any $a \in W$, the map

$$i_{*,a}: T_a(W) \to T_{i(a)}(V)$$

is injective. Two such pairs (W_1, i_1) and (W_2, i_2) are said to be equivalent if there is a C^k diffeomorphism h: $W_1 \to W_2$ such that $i_1 = i_2 \circ h$. This is an equivalence relation.

2.5.1 Definition. A *submanifold* of V is an equivalence class of pairs (W, i) under the equivalence defined above.

We shall often identify a submanifold with one of its representatives when no confusion is likely.

If $m = \dim W = \dim T_a(W)$, $a \in W$ and $n = \dim V = \dim T_{i(a)}(V)$, we clearly have $m \leqq n$.

2.5.2 Remark. Let $\dim W = m$ and $a \in W$. Then, by the rank theorem 2.2.11 there are coordinate systems (U, φ) at a and (U', φ') at $i(a)$ such that $\varphi' \circ i \circ \varphi^{-1}|\varphi(U)$ is the map

$$(x_1, \ldots, x_m) \mapsto (x_1, \ldots, x_m, 0, \ldots, 0).$$

It follows that if (U_1, ψ_1) is any system of coordinates at $a \in W$, $\psi_1(x) = (y_1, \ldots, y_m)$ then there exists a system of coordinates (U_2, ψ_2) at $i(a)$ such that if $\psi_2(u) = (u_1, \ldots, u_n)$, $u \in U_2$, we have $u_j \circ i = y_j$ on some neighbourhood of a in W for $j \leqq m$.

2.5.3 Proposition. Let W be a closed subset of the C^k manifold V of dimension n. Suppose that to each $a \in W$, there exists a coordinate system (U, φ) on V such that, if $\varphi(x) = (x_1, \ldots, x_n)$, we have

$$W \cap U = \{x \in U | x_{r+1} = \ldots = x_n = 0\};$$

here r is a fixed integer with $0 \leqq r \leqq n$. Then W carries a natural structure of a C^k manifold and the injection of W in V makes of W a submanifold of V.

The proof is very easy and is omitted.

2.5.4 REMARK. Definitions and remarks corresponding to 2.5.1–2.5.3 apply to real and complex analytic manifolds.

2.5.5 COROLLARY. Let V be a C^k (real, complex analytic) manifold and $f_1, \ldots, f_p$ $(1 \leqq p < n)$ be C^k (real analytic, holomorphic) functions on V. Let

$$W = \{x \in V | \ f_1(x) = \ldots = f_p(x) = 0\},$$

and suppose that $(df_1)_a, \ldots, (df_p)_a$ are linearly independent for all $a \in W$. Then (W, i), i being the injection, is a C^k (real, complex analytic) submanifold of V of dimension $n-p$. This is an immediate consequence of proposition 2.5.3 and the rank theorem 2.2.11; 2.4.9.

2.5.6 EXAMPLE. The n-sphere $S^n \subset R^{n+1}$ is defined by

$$S^n = \{x = (x_0, \ldots, x_n) \in R^{n+1} | x_0^2 + \ldots + x_n^2 = 1\}.$$

Since if $f(x) = x_0^2 + \ldots + x_n^2 - 1$, we have $(df)_a \neq 0$ for $a \in S^n$, S^n carries a natural real analytic structure by corollary 2.5.5.

In general, if (W, i) is a submanifold of V, i is not a homeomorphism of W onto $i(W)$ (with the topology induced from V). We now look for conditions on i in order that this be the case.

2.5.7 DEFINITION. Let V and W be locally compact Hausdorff spaces. A continuous map $f: V \to W$ is called *locally proper* if for every $w \in f(V)$, there is compact neighbourhood K of w in W such that $f^{-1}(K)$ is compact. f is called *proper* if, for any compact set K in W, $f^{-1}(K)$ is compact.

It is known that a proper map carries closed sets in V onto closed sets in W (see Bourbaki [1965]). It follows easily from this that a *locally proper map* f: $V \to W$ *is proper if and only if* $f(V)$ is *closed* in W. [Any proper map is, of course, locally proper.]

2.5.8 PROPOSITION. Let V be a C^k manifold and (W, i) a submanifold. Then the map $i: W \to i(W)$ is a homeomorphism of W onto $i(W)$ with the topology induced from V if and only if i is locally proper.

PROOF. Suppose first that i is locally proper. Then, given $a \in W$, there is a compact neighbourhood K of $i(a)$ in V such that $i^{-1}(K)$ is compact, hence a compact neighbourhood of a. Now

$$i: i^{-1}(K) \to K \cap i(W)$$

is a continuous bijective map of a compact Hausdorff space onto a Hausdorff space and so is a homeomorphism. Clearly this implies that i^{-1} is continuous.

Suppose now that $i: W \to i(W)$ is a homeomorphism. Then if $a \in W$ and C is a compact neighbourhood of a in W, then $i(C)$ is a compact neighbourhood of $i(a)$ in $i(W)$. Hence there is an open set D in V such that

$$i(a) \in D \cap i(W) \subset i(C).$$

Let K be a compact neighbourhood of $i(a)$, $K \subset D$. Clearly

$$K \cap i(W) \subset D \cap i(W) \subset i(C).$$

Since i is injective, this implies that $i^{-1}(K) \subset C$ and so $i^{-1}(K)$ is compact.

2.5.9 DEFINITION. A submanifold $\{(W, i)\}$ of V is called *closed* if, for some pair (W, i) representing it, the map $i: W \to V$ is proper.

Note that if (W', i') is equivalent with (W, i), and i is proper, then so is i'.

We give now an example of a submanifold which does not preserve topology. The example uses the following theorem due to Kronecker, which we do not prove. A beautiful and simple proof of a considerably stronger assertion is given by WEYL [1916].

2.5.10 KRONECKER'S THEOREM. Let $\lambda_1, \ldots, \lambda_n$ be real numbers which are linearly independent over the ring Z of integers. Let

$$T^n = S^1 \times \ldots \times S^1 = \{(e^{i\theta_1}, \ldots, e^{i\theta_n}) \mid \theta_j \in R\}$$

and let κ: $\mathsf{R} \to T^n$ be the map

$$t \mapsto (e^{i\lambda_1 t}, \ldots, e^{i\lambda_n t}).$$

Then $\kappa(\mathsf{R})$ is dense in T^n.

2.5.11 Example. The torus $T^n = S^1 \times \ldots \times S^1$, $n > 1$, is an analytic manifold of dimension n. Consider the map κ defined above. It is an injection, for if $\kappa(t_1) = \kappa(t_2)$, then

$$\lambda_j t_1 = \lambda_j t_2 + 2\pi m_j, \qquad m_j \in \mathsf{Z}, \qquad j = 1, \ldots, n.$$

If $t_1 \neq t_2$, and if $k_1, \ldots, k_n$ are integers, not all zero, with

$$\sum_{j=1}^{n} k_j m_j = 0,$$

we have

$$(t_1 - t_2) \sum_{j=1}^{n} k_j \lambda_j = 0,$$

contradicting our assumption that the λ_j aie independent over Z. [The k_j exist since $n > 1$.] Further, rank $(d\kappa)_a = 1$ foi any $a \in \mathsf{R}$. It follows that (T^n, κ) defines a submanifold. But κ is not locally proper. If it were, it would follow that any $x_0 \in \kappa(\mathsf{R}) = S$ has a neighbourhood U such that $S \cap U$ is closed in U (since a proper map carries closed sets onto closed sets). Since, by theorem 2.5.10, S is dense in T^n, so that $\bar{S} \cap U = U$, this would imply that S is open. But, since $n > 1$, S can contain no open subset of T^n (for example, because of lemma 1.4.3).

The next three propositions relate the structure of a C^k submanifold to that of the structure of the 'big' manifold.

2.5.12 Proposition. Let V be a C^k manifold and (W, i) a submanifold. Let M be any C^k manifold and f: $M \to W$ a continuous map. Then f is C^k if and only if $i \circ f$: $M \to V$ is C^k.

Proof. Let $a \in W$; choose coordinate systems (U, φ) at a, (U, φ') at $i(a)$ such that if

$$\varphi(x) = (x_1, \ldots, x_m), \qquad \varphi'(U) = (u_1, \ldots, u_n),$$

then $\varphi' \circ i \circ \varphi^{-1}|\varphi(U)$ is the map

$$(x_1, \ldots, x_m) \mapsto (x_1, \ldots, x_m, 0, \ldots, 0).$$

Then if

$$\varphi \circ f(y) = (y_1, \ldots, y_m),$$

($y \in f^{-1}(U)$ which is open, since f is continuous), we have

$$\varphi' \circ i \circ f(y) = (y_1, \ldots, y_m, 0, \ldots, 0).$$

The proposition asserts merely that

$$y \mapsto (y_1, \ldots, y_m, 0, \ldots, 0)$$

is C^k if and only if

$$y \mapsto (y_1, \ldots, y_m)$$

is C^k.

2.5.13 Proposition. Let V be a C^k manifold and (W, i) a C^k submanifold. A germ g_a of continuous functions at $a \in W$ is C^k, if and only if there is a germ G_b of C^k functions at $b = i(a)$ with $G_b \circ i = g_a$. Conversely, if i is a continuous injection of the C^k manifold W into the C^k manifold V with this property, the pair (W, i) is a submanifold.

Proof. Suppose that (W, i) is a C^k submanifold. Choose coordinates (U, φ) at $a \in W$, (U', φ') at $b = i(a)$ such that $\varphi(U)$, $\varphi'(U')$ are cubes in R^m, R^n respectively and $\varphi' \circ i \circ \varphi^{-1}|\varphi(U)$ is the map

$$(x_1, \ldots, x_m) \mapsto (x_1, \ldots, x_m, 0, \ldots, 0).$$

Then, if $g \in C^k(U)$ and $G \in C^k(U')$ is defined by

$$G \circ \varphi'^{-1}(x_1, \ldots, x_n) = g \circ \varphi^{-1}(x_1, \ldots, x_m),$$

we have clearly $G \circ i = g$. The first part of the proposition follows at once.

Suppose, conversely, that $i: W \to V$ is a continuous injection such that C^k germs at $a \in W$ are precisely the germs $G \circ i$ where G runs over the C^k germs at $b = i(a)$.

We first assert that i is a C^k map. Let $a \in W$ and $b = i(a)$, and let (U', φ') be a coordinate system at b, $\varphi'(u) = (u_1, \ldots, u_n)$. Then $\varphi' \circ i|i^{-1}\varphi'^{-1}(U')$ is the map $x \mapsto (x_1, \ldots, x_n)$, where $x_j = u_j \circ i$. Since the $u_j \in C^k(U')$, the $x_j \in C^k(\varphi'^{-1}(U'))$ by hypothesis; this means that i is a C^k map.

We now assert that if $a \in W$, $i(a) = b$, there are neighbourhoods D of a and D' of b and a C^k map $p: D' \to D$ such that $p \circ i =$ identity on D. Let (U, φ) be a coordinate system at a, $\varphi(x) = (x_1, \ldots, x_m)$. By hypothesis, if U is small enough, there exist C^k functions p_j in a neighbourhood U' of b such that $i(U) \subset U'$ and $p_j \circ i = x_j$. We suppose U, U' so small that there are coordinates φ' on U'. Clearly, if $q = (p_1, \ldots, p_m)$ we can choose a neighbourhood D' of a such that $p(D') \subset U$ where $p = \varphi^{-1} \circ q$. Clearly $p \circ i =$ identity in a neighbourhood of a; in particular, if D' is small enough, $p(D') = D$ is open in W and $p \circ i =$ identity in D.

We have $p_{*,b} \circ i_{*,a} =$ identity on $T_a(W)$ (by remark 2.2.1) so that, in particular $i_{*,a}$ is injective.

2.5.14 Proposition. Let V be a C^k manifold having a countable base and (W, i) a closed C^k submanifold (i.e., i: $W \to V$ is a proper map). Then, for any C^k function g on W, there is a C^k function G on V such that $g = G \circ i$.

Proof. Let $W' = i(W)$; W' is closed in V. Now, for each $a \in W'$, there is a neighbourhood U_a of a in V and a C^k function G_a in U_a such that $G_a \circ i = g$ on $i^{-1}(U_a)$; this follows from proposition 2.5.13 and the fact that i: $W \to W'$ is a homeomorphism. Consider the open covering $\{U_a, V - W'\}_{a \in W'}$ of V. By theorem 2.2.14, there is a partition of unity $\{\eta_a, \eta\}$ relative to this covering, i.e., C^k functions $\{\eta_a, \eta\}$ on V such that $\eta_a, \eta \geqq 0$, supp $(\eta_a) \subset U_a$, supp $(\eta) \subset V - W'$, whose supports form a locally finite family and

$$\eta(x) + \sum_{a \in W'} \eta_a(x) = 1.$$

Define $h_a \in C^k(V)$ by

$$h_a(x) = \begin{cases} \eta_a(x) G_a(x) & x \in U_a \\ 0 & x \notin U_a \end{cases}$$

then $\{\text{supp}\,(h_a)\}$ is locally finite, so that

$$G = \sum_{a \in W'} h_a \in C^k(V).$$

If $w \in W$ and

$$A = \{a \in W' | \quad i(w) \in U_a\},$$

we have

$$\sum_{a \in A} \eta_a(i(w)) = \eta(i(w)) + \sum_{a \in W'} \eta_a(i(w)) = 1.$$

Hence

$$G \circ i(w) = \sum_{a \in A} G_a(i(w))\eta_a(i(w)) = g(w) \sum_{a \in A} \eta_a(i(w)) = g(w).$$

2.5.15 REMARKS. Propositions 2.5.12; 2.5.13 and their proofs remain valid for real or complex manifolds with the corresponding mappings and functions. Proposition 2.5.14 is true for real analytic manifolds (with countable base) and analytic functions. This is however very difficult to prove; see CARTAN [1957] and GRAUERT [1958]. The corresponding statement for complex manifolds (and holomorphic functions) is false in general. A very important case in which it is valid (due to OKA [1936]) will be dealt with later (see theorem 2.14.9).

2.5.16 REMARK. If V is a C^k manifold, (W, i) a closed submanifold and ω is a p-form on V, we denote by $\omega | W$ the form $i^*(\omega)$ on W.

§ 2.6 Exterior differentiation

Let V be a C^k manifold, $k \geqq 2$, and let $A^p(V, r)$, $p > 0$, $0 \leqq r < k$ denote the set of all C^r differential forms of degree p on V. If $p = 0$, then $A^0(V, r)$ for $0 \leqq r \leqq k$ denotes the set of all C^r functions on V. The results that we discuss now are true of both real- and complex-valued functions and forms.

2.6.1 DEFINITION. An *exterior derivative* $d = d_V$ is an assignment, to each pair (p, r) with $1 \leqq r < k$ if $p > 0$, and $1 \leqq r \leqq k$ if $p = 0$, of a map $d\colon A^p(V, r) \to A^{p+1}(V, r-1)$ which satisfies the following conditions.

(a) d is R (or C) linear for each p and r.

(b) If $f \in A^0(V, r)$, $1 \leqq r \leqq k$, df is the 1-form ω defined by $\omega(a) = (df)_a =$ image of f in $T_a^*(V)$.

(c) If $f \in A^0(V, r)$, $2 \leqq r \leqq k$, we have $d(df) = 0$.

(d) If $\omega_1 \in A^p(V, r)$, $\omega_2 \in A^q(V, r)$, then
$$d(\omega_1 \wedge \omega_2) = d\omega_1 \wedge \omega_2 + (-1)^p \omega_1 \wedge d\omega_2.$$

Before we proceed to prove the existence and uniqueness of the exterior derivative, we deduce some consequences of (a)–(d) above.

2.6.2 PROPOSITION. d is a local operator. In other words, if $\omega \in A^p(V, r)$ and $\omega | U = 0$ for some open set U on V, then $d\omega | U = 0$.

PROOF. Let $a \in U$ and let f be a C^k function on V which $= 0$ in a neighbourhood of a and $= 1$ in a neighbourhood of $V - U$ (corollary 2.2.15). Then $\omega = f\omega$. Hence

$$d\omega = (df) \wedge \omega + f \wedge (d\omega).$$

Now $f = 0$ in a neighbourhood of a and $\omega(a) = 0$. Hence $(d\omega)(a) = 0$. Since $a \in U$ is arbitrary, this proves the proposition.

Note that the existence of f above does not require the assumption that V have a countable base, although corollary 2.2.15 does. In fact, if U' is a relatively compact neighbourhood of a, there exists, by corollary 2.2.15, a C^k function $= 0$ near a and $= 1$ in a neighbourhood of the boundary $\partial U'$ of U'; we can simply define f to be 1 on $V - U'$.

2.6.3 PROPOSITION. Let $k \geqq 3$, $r \geqq 2$. Then $d^2 = 0$ on $A^p(V, r)$.

PROOF. We may suppose that $p \geqq 1$. Let $a \in V$. We remark that if $\omega \in A^p(V, r)$, we can find a p-form ω' which is a finite linear combination of forms of the type

2.6.4 $$g \cdot dg_1 \wedge \ldots \wedge dg_p, \qquad g \in C^r(V), \quad g_j \in C^k(V)$$

such that $\omega - \omega' = 0$ in a neighbourhood of a. In fact, if (V, φ) is a coordinate system at a, $\varphi(x) = (x_1, \ldots, x_n)$ and

$$\omega(x) = \sum_J f_J(x) dx_{j_1} \wedge \ldots \wedge dx_{j_p}, \quad J = (j_1, \ldots, j_p), \quad j_1 < \ldots < j_p,$$

we can take for ω' the form obtained by replacing f_J by ηf_J and x_j by ηx_j, where $\eta \in C^k(V)$ and $\eta = 1$ in a neighbourhood of a and has compact support contained in U. Hence, by proposition 2.6.2, it suffices to prove that $d^2\omega = 0$, if ω has the form (2.6.4). Now, by induction on p and conditions 2.6.1.c,d, we have

$$d(dg_1 \wedge \ldots \wedge dg_p) = 0.$$

Hence, condition 2.6.1.d gives

$$d^2(gdg_1 \wedge \ldots \wedge dg_p) = d^2(g) \wedge dg_1 \wedge \ldots \wedge dg_p = 0$$

(the last step because of condition 2.6.1.c).

2.6.5 THEOREM. If V is a C^k manifold, $k \geqq 2$, then the exterior derivative exists and is unique.

PROOF. *Uniqueness*: Let d_1, d_2 be two exterior derivatives. By condition 2.6.1.b, $d_1 f = d_2 f$ if $f \in C^r(V)$; both are, moreover, local operators. It is therefore sufficient to show that $d_1\omega = d_2\omega$ in the neighbourhood of any given point $a \in V$. Now, ω is equal, in the neighbourhood of a, to a finite linear combination of forms of the type (2.6.4), Hence, by proposition 2.6.2, it is sufficient to prove that

$$d_1(gdg_1 \wedge \ldots \wedge dg_p) = d_2(gdg_1 \wedge \ldots \wedge dg_p).$$

But again, by induction on p and conditions 2.6.1.c and d, we have

$$d_j(gdg_1 \wedge \ldots \wedge dg_p) = dg \wedge dg_1 \wedge \ldots \wedge dg_p, \quad j = 1, 2.$$

This proves uniqueness.

Existence: Because of the uniqueness proved above, and proposition 2.6.2, it is sufficient to prove the existence of the exterior derivative when V is C^k diffeomorphic to an open set Ω in R^n. Let $\varphi : V \to \Omega$ be a C^k diffeomorphism and $\varphi(x) = (x_1, \ldots, x_n)$. Any $\omega \in A^p(V, r)$ can be uniquely written in the form

$$\omega(x) = \sum_J f_J(x) dx_{j_1} \wedge \ldots \wedge dx_{j_p}, \quad J = (j_1, \ldots, j_p),$$
$$j_1 < \ldots < j_p, \quad f_J \in C^r(V).$$

We define

$$d\omega = \sum_J df_J \wedge dx_{j_1} \wedge \ldots \wedge dx_{j_p},$$

where df_J is defined by condition 2.6.1.b. Trivially this operator satisfies conditions 2.6.1.a,b. As for condition 2.6.1.c, we have, by proposition 2.1.11′,

$$(df)(x) = \sum_{\nu=1}^{n} \left(\frac{\partial}{\partial x_\nu}\right)_x f \cdot dx_\nu,$$

$$(d^2 f)(x) = \sum_{\nu=1}^{n} \left\{ \sum_{\mu=1}^{n} \left(\frac{\partial}{\partial x_\mu}\right)_x \left[\left(\frac{\partial}{\partial x_\nu}\right)_x f \right] dx_\mu \right\} \wedge dx_\nu$$

$$= \sum_{\mu<\nu} \left[\left(\frac{\partial}{\partial x_\mu}\right), \left(\frac{\partial}{\partial x_\nu}\right) \right] f \cdot dx_\mu \wedge dx_\nu = 0,$$

since the Poisson bracket of the vector fields $(\partial/\partial x_\mu)$, $(\partial/\partial x_\nu)$ is zero.

To prove condition 2.6.1.d, we can assume that

$$\omega_1 = f_1 \, dx_{j_1} \wedge \ldots \wedge dx_{j_p}, \qquad \omega_2 = f_2 \, dx_{k_1} \wedge \ldots \wedge dx_{k_q};$$

we write $\omega_1 = f_1 \, dx_J$ and $\omega_2 = f_2 \, dx_K$. Now we have

$$\omega_1 \wedge \omega_2 = (f_1 f_2) dx_J \wedge dx_K .$$

Clearly, by definition,

$$d(\omega_1 \wedge \omega_2) = d(f_1 f_2) \wedge dx_J \wedge dx_K = \{(df_1) f_2 + f_1 (df_2)\} \wedge dx_J \wedge dx_K$$
$$= df_1 \wedge dx_J \wedge f_2 \, dx_K + (-1)^p f_1 \, dx_J \wedge (df_2) \wedge dx_K;$$

we are using the fact that

$$d(f_1 f_2) = (df_1) f_2 + f_1 (df_2)$$

and the fact that if E is a vector space, we have

$$e_0 \wedge e_1 \wedge \ldots \wedge e_p = (-1)^p e_1 \wedge \ldots \wedge e_p \wedge e_0, \quad e_j \in E, \quad 0 \leqq j \leqq p.$$

This clearly proves condition 2.6.1.d and with it theorem 2.6.1.

We have already remarked that $T_a^*(V)$ is the dual of $T_a(V)$ (proposition 2.1.11). Hence $\wedge^p T_a^*(V)$ is the dual of $\wedge^p T_a(V)$, so that any $\omega_a \in \wedge^p T_a^*(V)$ can be looked upon as an alternate multilinear function on $\oplus_{\nu=1}^{p} T_a(V)$, and any p-form ω gives rise to an alternate multilinear map of the space

$$\mathfrak{X}^p = \bigoplus_{\nu=1}^{p} \mathfrak{X},$$

where $\mathfrak{X}$ is the space of vector fields C^{k-1} on V (see proposition 2.4.3), which is multilinear over the ring of C^{k-1} functions.

2.6.6 PROPOSITION. If ω is a C^{k-1} p-form and $X_1, \ldots, X_{p+1} \in \mathfrak{X}$, then the map $d\omega$: $\mathfrak{X}^{p+1} \to R_{k-2}$(where R_{k-2} is the ring $C^{k-2}(V)$) is given by

$$(d\omega)(X_1, \ldots, X_{p+1}) = \sum_{\nu=1}^{p+1} (-1)^{\nu+1} X_\nu(\omega(X_1, \ldots, \hat{X}_\nu, \ldots, X_{p+1}))$$
$$+ \sum_{\mu<\nu} (-1)^{\mu+\nu} \omega([X_\mu, X_\nu], X_1, \ldots, \hat{X}_\mu, \ldots, \hat{X}_\nu, \ldots, X_{p+1});$$

here the hat $\wedge$ over a term X_ν means that that term is to be omitted.

Since the computations involved in the proof of this proposition are cumbersome, we prove it only in the case $p=1$; a proof in the general case which is obtained from a quite different point of view, is given in KOSZUL [1960], where, in fact, this formula is taken as *definition* of exterior derivation.

PROOF OF PROPOSITION 2.6.6 WHEN p = 1. It is sufficient to prove this formula in the neighbourhood of any point; thus we may suppose that $\omega = f_1 df_2$ where $f_1 \in C^{k-1}(V)$, $f_2 \in C^k(V)$. Further we have $d\omega = (df_1) \wedge (df_2)$. This gives

$$(df_1 \wedge df_2)(X_1, X_2) = \det \begin{pmatrix} (df_1)(X_1) & (df_1)(X_2) \\ (df_2)(X_1) & (df_2)(X_2) \end{pmatrix}$$
$$= X_1(f_1)X_2(f_2) - X_1(f_2)X_2(f_1)$$
$$= X_1[f_1 X_2(f_2)] - X_2[f_1 X_1(f_2)] - f_1 X_1(X_2(f_2)) + f_1 X_2(X_1(f_2))$$
$$= X_1(\omega(X_2)) - X_2(\omega(X_1)) - \omega([X_1, X_2]),$$

which is the required formula.

2.6.7 PROPOSITION. Let V, W be C^k manifolds and f: $V \to W$ a C^k map. Then if ω is a p-form on W, we have

2.6.8 $$d_V f^*(\omega) = f^*(d_W \omega);$$

here d_V, d_W denote, respectively, the exterior derivatives on V and W.

PROOF. It is sufficient to prove this result when W is C^k diffeomorphic to an open set in R^m. Moreover, since

$$f^*: A(W) \to A(V)$$

(space of all differential forms) is an algebra homomorphism, it is sufficient, because of condition 2.6.1.d, to prove (2.6.8.) for a system of elements of $A(W)$ which generate it as an algebra. Since W is diffeomorphic to an open set in R^n, such a system is given by the elements $\{h, dg\}$, where $h \in C^{k-1}(W)$ and $g \in C^k(W)$.

By definition of f_a^*,

$$f_a^*(dh)_{f(a)} = d(h \circ f)_a,$$

which gives (2.6.8) if $\omega = h$. If $\omega = dg$,

$$f^*(d_W \omega) = 0 \quad \text{and} \quad d_V f^*(\omega) = d_V(d_V(g \circ f)) = 0.$$

This proves the result.

2.6.9 DEFINITION. A differential form ω on the C^k manifold V is called *closed* if $d\omega = 0$. It is called *exact* if there is a differential form ω' such that $d\omega' = \omega$; here ω, ω' are C^{k-1}.

If V is a C^∞ manifold, we denote the set of closed p-forms by $Z^p(V)$, that of exact p-forms by $B^p(V)$. By proposition 2.6.3, we have $B^p(V) \subset Z^p(V)$. The quotients

$$H^p(V) = Z^p(V)/B^p(V)$$

are called the de Rham cohomology groups of V. A basic theorem, due to de Rham, asserts that these groups are *topological invariants* of V, i.e., that *homeomorphic* C^∞ manifolds have isomorphic de Rham groups. For a proof, see, e.g., WEIL [1952].

We look now at the exterior derivative on a complex manifold. Let V be a complex manifold, $a \in V$ and $\mathscr{E}^*_{p,q}(V, a)$ the space of covectors of type (p, q) at a. We denote by $\mathscr{A}^{p,q}(V)$ the space of all C^∞ forms of type (p, q) on V;

$$\mathscr{A}(V) = \sum_{p,q \geq 0} \mathscr{A}^{p,q}(V)$$

is then the space of all complex-valued differential forms on V. As before, we can define the exterior derivative d: $\mathscr{A}(V) \to \mathscr{A}(V)$ with the properties 2.6.1.a-d. In fact

$$\mathscr{A}(V) = A(V) \otimes_R \mathbf{C},$$

and the operator d defined earlier extends to $\mathscr{A}(V)$.

If $f \in C^\infty(V, \mathbf{C})$, and $a \in V$, then

$$(df)_a \in \mathfrak{T}_a^*(V) = \mathscr{E}_{1,0}^* \oplus \mathscr{E}_{0,1}^*.$$

Hence

$$(df)_a = (\partial f)_a + (\bar{\partial} f)_a$$

where

$$(\partial f)_a \in \mathscr{E}_{1,0}^*(V, a), \qquad (\bar{\partial} f)_a \in \mathscr{E}_{0,1}^*(V, a).$$

Hence the exterior derivative

$$df = \partial f + \bar{\partial} f,$$

where

$$\partial f \in \mathscr{A}^{1,0}(V), \qquad \bar{\partial} f \in \mathscr{A}^{0,1}(V).$$

We assert now that

2.6.10 $$d\mathscr{A}^{p,q}(V) \subset \mathscr{A}^{p+1,q}(V) + \mathscr{A}^{p,q+1}(V).$$

This is a local question, so that we may suppose that V is a coordinate neighbourhood U. Let $\varphi(z) = (z_1, \ldots, z_n)$ be complex coordinates in U. Then, any element of $A^{p,q}(U)$ is a finite linear combination of elements of the form

$$\omega' = f dz_{j_1} \wedge \ldots \wedge dz_{j_p} \wedge d\bar{z}_{k_1} \wedge \ldots \wedge d\bar{z}_{k_q} \equiv f dz_J \wedge d\bar{z}_K.$$

But

$$d\omega' = (df) \wedge dz_J \wedge d\bar{z}_K = \partial f \wedge dz_J \wedge d\bar{z}_K + (-1)^p dz_J \wedge \bar{\partial} f \wedge d\bar{z}_K$$

and this clearly lies in

$$\mathscr{A}^{p+1,q}(U) + \mathscr{A}^{p,q+1}(U).$$

Thus, if $\omega \in \mathscr{A}^{p,q}(V)$, we define $\partial\omega$, $\bar{\partial}\omega$ by the conditions

2.6.11 $$d\omega = \partial\omega + \bar{\partial}\omega, \qquad \partial\omega \in \mathscr{A}^{p+1,q}(V), \qquad \bar{\partial}\omega \in \mathscr{A}^{p,q+1}(V).$$

Clearly then $\partial, \bar{\partial}$ extend to $\mathbf{C}$ linear operators on $\mathscr{A}(V)$. Since $\mathscr{A}(V)$ is the *direct* sum $\sum_{p,q \geqq 0} \mathscr{A}^{p,q}(V)$, we see at once that the fact that $d^2 = 0$ is equivalent with

2.6.12 $$\partial^2 = 0, \qquad \bar{\partial}^2 = 0, \qquad \partial\bar{\partial} + \bar{\partial}\partial = 0.$$

Further, if $e \leftrightarrow \bar{e}$ denotes the conjugation $\mathscr{E}_{1,0}^* \to \mathscr{E}_{0,1}^*$ defined in § 4, we have

$$\overline{\partial f} = \bar{\partial}(\bar{f}), \qquad f \in C^\infty(V, \mathbf{C}).$$

Let V, V' be complex manifolds and f: $V \to V'$ a holomorphic map. Let ω' be a form of type (p, q) on V'. Now by proposition 2.6.7

$$d_V f^*(\omega') = f^*(d_W \omega').$$

Hence

$$\partial f^*(\omega') + \bar{\partial} f^*(\omega') = f^*(\partial\omega') + f^*(\bar{\partial}\omega').$$

Now, by proposition 2.4.12, the pull back of a form of type (p, q) by a holomorphic map is again of type (p, q). This and the above equation imply

2.6.13 $$\partial f^* = f^*\partial, \qquad \bar{\partial} f^* = f^*\bar{\partial},$$

when f: $V \to V'$ is holomorphic.

2.6.14 DEFINITION. A differential form ω of degree p on the complex manifold is *holomorphic* if ω is of type $(p, 0)$ and $\bar{\partial}\omega = 0$.

2.6.15 REMARK. A function $f \in C^\infty(V, C)$ represents a holomorphic form of degree 0 if and only if it is a holomorphic function. A form ω of type $(p, 0)$ is holomorphic if and only if, for every complex coordinate system (U, φ), $(\varphi(z) = (z_1, \ldots, z_n))$ ω has the expression:

$$\omega = \sum_J f_J dz_{j_1} \wedge \ldots \wedge dz_{j_p}, \qquad J = (j_1, \ldots, j_p), \quad j_1 < \ldots < j_p$$

in which all the f_J are holomorphic functions. These are immediate corollaries of the next remark.

2.6.16 REMARK. Note that in proving (2.6.10), we have shown that if in local coordinates, a form ω of type (p, q) has the expression

$$\omega = \sum_{J,K} f_{JK} dz_J \wedge d\bar{z}_K, \qquad J = (j_1, \ldots, j_p), \qquad K = (k_1, \ldots, k_q),$$

$$j_1 < \ldots < j_p, \qquad k_1 < \ldots < k_q, \qquad dz_J = dz_{j_1} \wedge \ldots \wedge dz_{j_p},$$
$$d\bar{z}_K = d\bar{z}_{k_1} \wedge \ldots \wedge d\bar{z}_{k_q},$$

then

$$\begin{cases} \partial\omega = \sum_{J,K} \sum_{\nu} \left(\dfrac{\partial}{\partial z_\nu}\right) f_{JK} dz_\nu \wedge dz_J \wedge d\bar{z}_K \\ \bar{\partial}\omega = \sum_{J,K} \sum_{\nu} (-1)^p \left(\dfrac{\partial}{\partial \bar{z}_\nu}\right) f_{JK} dz_J \wedge d\bar{z}_\nu \wedge d\bar{z}_K. \end{cases}$$

§ 2.7 Orientation

2.7.1 DEFINITION. Let V be a C^k manifold, $k \geqq 1$, of dimension n, then is V said to be *orientable* if there is on V a continuous n-form ω which is nowhere zero.

If V is connected and ω_1, ω_2 are two n-forms which never vanish, there is a continuous function f such that $\omega_1 = f\omega_2$; moreover f is either everywhere > 0 or everywhere < 0. If $f(x) > 0$ for all x, we call ω_1 and ω_2 equivalent; an *orientation* of V is a choice of one class of equivalent n-forms. There are two possible orientations of an orientable connected manifold. Let V be a C^k manifold and (U_i, φ_i), (U_j, φ_j) be two coordinate systems on V. Then

$$\varphi_i \circ \varphi_j^{-1} \colon \varphi_j(U_i \cap U_j) \to R^n$$

is a C^k map and we set

$$d_{ij}(x) = \det \{d(\varphi_i \circ \varphi_j^{-1})(\varphi_j(x))\}.$$

2.7.2 PROPOSITION. V is orientable if and only if we can find coordinate systems $\{(U_i, \varphi_i)\}_{i \in \mathscr{I}}$ such that $V = \bigcup U_i$ and $d_{ij}(x) > 0$ if $x \in U_i \cap U_j$.

PROOF. We may suppose that V is connected. Let ω be a continuous n-form which is nowhere zero. If (U_a, φ_a) is a coordinate system, $\varphi_a(x) = (x_1, \ldots, x_n)$, we set $\omega_a = dx_1 \wedge \ldots \wedge dx_n$; this is an n-form on U_a without zeros. For any point $a \in V$, we can find a coordinate system (U_a, φ_a) with $a \in U_a$ such that $\omega = g_a \omega_a$, where g_a is a continuous function with $g_a(x) > 0$ for $x \in U_a$ (replacing $\varphi_a(x) = (x_1, \ldots, x_{n-1}, x_n)$ by $(x_1, \ldots, x_{n-1}, -x_n)$ if necessary). Moreover, we have, on $U_a \cap U_b$,

$$\omega_a = d_{ab}\, \omega_b\,.$$

Hence $d_{ab} = g_b/g_a > 0$ on $U_a \cap U_b$.

Suppose conversely that $\{U_i, \varphi_i\}$ is a family of coordinate systems with $V = \bigcup U_i$, $d_{ij} > 0$ on $U_i \cap U_j$. Define, as above

$$\omega_i = dx_1 \wedge \ldots \wedge dx_n$$

if $\varphi_i(x) = (x_1, \ldots, x_n)$. Let $\{\eta_i\}$ be a C^k partition of unity relative to $\{U_i\}$, and define

$$\omega = \sum \eta_i \omega_i\,.$$

Clearly ω is a C^{k-1} form. Moreover, if $a \in V$, and I denotes the set of i such that $a \in \operatorname{supp}(\eta_i)$ (which is a non-empty finite set) we have

$$\omega(a) = \sum_{i \in I} \eta_i(a)\omega_i(a) = \{\sum_{i \in I} \eta_i(a) d_{ii_0}(a)\}\omega_{i_0}(a),$$

where $i_0 \in I$ is fixed. Since

$$\sum_{i \in I} \eta_i(a) = 1$$

and $d_{ii_0}(a) > 0$ for any i, and $\eta_i \geqq 0$, we conclude that $\omega(a) \neq 0$.

2.7.3 REMARK. The above proof shows that on an orientable C^k manifold, there exists a C^{k-1} form without zeros.

We now consider the bundle

$$\wedge^n T^*(V) = \bigcup_{a \in V} \wedge^n T_a^*(V)$$

of n-forms on V, and the open set

$H = \{\xi \in \wedge^n T^*(V) | \xi \in \wedge^n T_a^*(V)$ and is not the zero in $\wedge^n T_a^*(V)\}$. Let $p: H \to V$ be the map $p(\xi) = a$ if $\xi \in \wedge^n T_a^*(V)$. We define on H an equivalence relation by: $\xi_1 \sim \xi_2$ if $p(\xi_1) = p(\xi_2)$ and $\xi_1 = \lambda\xi_2$ where $\lambda > 0$. Let $\tilde{V}$ be the quotient of H by this equivalence relation. It is easily seen that $\tilde{V}$ is a Hausdorff space. Further, the map $p: H \to V$ induces a continous map $\pi: \tilde{V} \to V$.

2.7.4 PROPOSITION. The map $\pi: \tilde{V} \to V$ is a 2-sheeted covering of V, i.e., π is a covering, and $\pi^{-1}(a)$ consists of two points for any $a \in V$.

PROOF. Let $a \in V$ and (U, φ) be a coordinate system with $a \in U$ in which U is connected. Let ω be the nowhere zero n-form on U defined by

$$\omega = dx_1 \wedge \ldots \wedge dx_n, \qquad \text{if} \quad \varphi(x) = (x_1, \ldots, x_n).$$

Let $\tilde{U}_1, \tilde{U}_2$ be the set of $\tilde{x} \in \tilde{V}$ such that $\pi(\tilde{x}) = x \in U$ and $\omega(x) \in \tilde{x}$ ($-\omega(x) \in \tilde{x}$ respectively). It is trivial that $\pi^{-1}(U) = \tilde{U}_1 \cup \tilde{U}_2$ and that $\pi | \tilde{U}_j$ is a homeomorphism onto U. This proves the result.

2.7.5 PROPOSITION. Let V be a connected C^k manifold. Then V is orientable if and only if $\tilde{V}$ is not connected.

PROOF. Let V be orientable and ω a nowhere zero n-form on V. ω is a continuous map $V \to H$, hence induces a continuous map $\bar{\omega}: V \to \tilde{V}$. Clearly $\pi \circ \bar{\omega} =$ identity, so that $\tilde{V}$ is not connected.

If $\tilde{V}$ is not connected, let $\tilde{W}$ be a connected component of $\tilde{V}$. Clearly $\pi | \tilde{W} \to V$ is a covering. Further, there is an $\tilde{x}_0 \in \tilde{V}$, $\tilde{x}_0 \notin \tilde{W}$. Hence

$$\{\tilde{x} \in \tilde{W} | \pi(\tilde{x}) = \pi(\tilde{x}_0)\}$$

contains just one element. Since $\pi | \tilde{W}$ is a covering, this implies that $\pi: \tilde{W} \to V$ is a homeomorphism.

Let (U_i, φ_i) be now a family of coordinate systems such that if we set $\omega_i = dx_1 \wedge \ldots \wedge dx_n$ $(\varphi_i(x) = (x_1, \ldots, x_n))$, then we have $\omega_i(x) \in \tilde{W}$ for $x \in U_i$ and $V = \bigcup U_i$ {such a family exists; if $\omega_i(x_0) \notin \tilde{W}$ for some x_0, replace $(x_1, \ldots, x_{n-1}, x_n)$ by $(x_1, \ldots, x_{n-1}, -x_n)$}. If $\omega_i = d_{ij}\omega_j$ on $U_i \cap U_j$, we find, by definition of the equivalence relation on H, that $d_{ij} > 0$ on $U_i \cap U_j$. Hence, by proposition 2.7.2, V is orientable.

2.7.6 COROLLARY. A connected, simply connected manifold is orientable; this is an immediate consequence of proposition 2.7.5.
Clearly $\tilde{V}$ carries a natural C^k structure (example 2.1.7.b). It can be proved that there is always (even when V is not orientable) an n-form $\tilde{\omega}$ such that for any $\xi \in \tilde{x} \in \tilde{V}$ we have $\tilde{\omega}(\tilde{x}) = \lambda \pi_x^*(\xi)$, $x = \pi(\tilde{x})$ where $\lambda > 0$. It follows that $\tilde{V}$ is *always orientable.*

§ 2.8 Manifolds with boundary

Let $R_+^n = \{(x_1, \ldots, x_n) \in R^n | x_1 \geqq 0\}$. Let Ω be an open subset of R_+^n. A function f on Ω is said to be of class C^k if there is an open set Ω' in R^n and an $F \in C^k(\Omega')$ such that $F|\Omega = f$; C^k mappings of Ω into R^m are similarly defined.

2.8.1 DEFINITION. Let V be a Hausdorff topological space. Suppose given a family of pairs $\{(U_i, \varphi_i)\}$ where U_i is open in V and φ_i is a homeomorphism of U_i on to an open set in R_+^n so that the map

$\varphi_i \circ \varphi_j^{-1}|\varphi_j(U_i \cap U_j)$ is C^k for all i, j. We say that this family defines the structure of a C^k manifold with boundary on V. A C^k structure on V is then a maximal family $\{(U_i, \varphi_i)\}$ with the above property.

For a C^k manifold V with boundary ($k \geqq 1$), one can define C^k functions and germs, tangent vectors, differential forms, orientation and so on in the same way as for ordinary C^k manifolds. We suppose that V is a C^k manifold with boundary with a countable base.

We consider elements $f \in C_{a,k}$, $a \in V$. We write $f \geqq 0$ if $f(x) \geqq 0$ for all x sufficiently near a (for some representative of the germ f).

2.8.2 DEFINITION. Let V be a C^k manifold with boundary. We say that a point $a \in V$ is an *interior point* if there is a coordinate system (U, φ) with $a \in U$, such that $\varphi(U)$ is an open set in R^n (i.e., $\varphi(U)$ does not meet the set $(t_1 = 0)$ in $R^n_+ = \{(t_1, \ldots, t_n) \in R^n | \, t_1 \geqq 0\}$). A point is called a *boundary point* if it is not an interior point, and we denote by ∂V the set of boundary points of V.

Remark that $a \in V$ belongs to ∂V if and only if there is a coordinate system (U, φ), $a \in U$, such that $\varphi(U) \subset R^n_+$ and $\varphi(a) = 0$. With respect to such a coordinate system, we have

$$\partial V \cap U = \{x \in U | \, x_1 = 0, \quad \text{where} \quad \varphi(x) = (x_1, \ldots, x_n)\}.$$

Clearly the map $(x_1, \ldots, x_n) \mapsto (x_2, \ldots, x_n)$ of R^n onto R^{n-1} induces a homeomorphism, which we denote by ψ, of $\partial V \cap U$ onto an open set in R^{n-1}. Furthermore, if (U, φ), (U', φ') are two coordinate systems on V (such that $U \cap U' \cap \partial V \neq \emptyset$), and if ψ, ψ' denote the corresponding homeomorphisms of $\partial V \cap U$, $\partial V \cap U'$ respectively, then clearly

$$\psi' \circ \psi^{-1}|\psi(U \cap U' \cap \partial V)$$

is the restriction to the set $t_1 = 0$ of the C^k map

$$\varphi' \circ \varphi^{-1}|\varphi(U \cap U')$$

and so is itself a C^k map. Hence we obtain:

2.8.3 PROPOSITION. The boundary ∂V of a C^k manifold V with boundary carries a natural structure of a C^k manifold (without boundary).

Moreover, it is seen at once that ∂V is a closed C^k submanifold of V (the natural injection $\partial V \to V$ makes it so), and we look upon the tangent space $T_a(\partial V)$ to ∂V at $a \in \partial V$ as a subspace of $T_a(V)$.

2.8.4 DEFINITION. Let V be a C^k manifold with boundary, and let $a \in V$. A tangent vector $X \in T_a(V)$ is called *positive* (or an inner normal to ∂V) if for any $f \in \boldsymbol{m}_{a,k}$, $f \geqq 0$, we have $X(f) \geqq 0$ and if there is at least one $f \in \boldsymbol{m}_{a,k}$, $f \geqq 0$, for which $X(f) > 0$.

2.8.5 REMARK. Let a be an interior point of V. Then, we assert that there exists no positive tangent vector at a. In fact if $f \in \boldsymbol{m}_{a,k}$, $f \geqq 0$, then f has a local minimum at a. From the fact that the first partial derivatives of a C^1 function in an open set in R^n with a local minimum vanish at this point, we conclude that $X(f) = 0$ for any $X \in T_a(V)$.

2.8.6 PROPOSITION. Let $a \in \partial V$ and let (U, φ) be a coordinate system at a such that $\varphi(U) \subset R^n_+$, $\varphi(a) = 0$. Let $\varphi(x) = (x_1, \ldots, x_n)$ and, for, $X \in T_a(V)$ let

$$X = \sum_{\nu=1}^{n} c_\nu \left(\frac{\partial}{\partial x_\nu}\right)_a .$$

Then X is positive if and only if $c_1 > 0$.

PROOF. If $f \in \boldsymbol{m}_{a,k}$, $f \geqq 0$, the function $g \in \boldsymbol{m}_{0,k}(R^{n-1})$ defined by

$$g(x_2, \ldots, x_n) = f(0, x_2, \ldots, x_n) \qquad [\varphi(x) = (x_1, \ldots, x_n)]$$

has a local minimum at 0. Hence $(\partial g/\partial x_\nu)_0 = 0$ for $\nu \geqq 2$. It follows that $X(f) = c_1(\partial/\partial x_1)_a f$. Moreover

$$\left(\frac{\partial}{\partial x_1}\right)_a f = \lim_{h \to +0} h^{-1}\{f(h, 0, \ldots, 0) - f(0, \ldots, 0)\}$$
$$= \lim_{h \to +0} h^{-1} f(h, 0, \ldots, 0) \geqq 0.$$

Further

$$\left(\frac{\partial}{\partial x_1}\right)_a x_1 = 1.$$

The proposition follows at once.

2.8.7 PROPOSITION. If $a \in \partial V$ and X_1, X_2 are positive tangent vectors at a, then there is a $\lambda > 0$ so that $X_1 - \lambda X_2 \in T_a(\partial V)$.

PROOF. If (U, φ) is a coordinate system at a, $\varphi(a) = 0$, let

$$X_j = \sum_{\nu=1}^{n} c_\nu^{(j)} \left(\frac{\partial}{\partial x_\nu}\right)_a.$$

Choose $\lambda > 0$ so that

$$c_1^{(1)} - \lambda c_1^{(2)} = 0$$

(possible by proposition 2.8.6). Then $X_1 - \lambda X_2 \in T_a(\partial V)$.

2.8.8 DEFINITION. An element $\omega \in T_a^*(V)$, $a \in \partial V$, is called *positive* if $\omega(X) > 0$ for all positive $X \in T_a(V)$.

In terms of local coordinates, (U, φ), $\varphi(a) = 0$, $\varphi(x) = (x_1, \ldots, x_n)$, one sees that ω has the form $c(dx_1)_a$, $c > 0$.

2.8.9 PROPOSITION. An orientation of V induces a natural orientation of ∂V.

PROOF. Let ω be an n-form on V which is nowhere zero. We cover ∂V by coordinates systems, (U_i', φ_i') such that if

$$\omega_i' = dx_2 \wedge \ldots \wedge dx_n \quad [\varphi_i'(x) = (x_2, \ldots, x_n)]$$

then, for any $y \in U_i'$ and any positive $e \in T_y^*(V)$ (see definition 2.8.8), we have

$$e \wedge \omega_i'(y) = -\lambda \omega(y)$$

with $\lambda > 0$ [we are looking upon $\wedge^{n-1} T_y^*(\partial V)$ as a subspace of $\wedge^{n-1} T_y^*(V)$]. It is immediate that $\omega_i' = d_{ij}' \omega_j'$ with $d_{ij}' > 0$ in $U_i' \cap U_j'$. One concludes from the proof of proposition 2.7.2 that there is an $(n-1)$-form ω' on ∂V such that $\omega_i' = g_i' \omega'$, where $g_i' > 0$ in U_i'.

2.8.10 REMARK. If D is an open set on the C^k manifold V such that for any $a \in \bar{D} - D$, there exists a neighbourhood U in V and a C^k function g in U with

$$(dg)_a \neq 0, \qquad D \cap U = \{x \in U | g(x) > 0\},$$

then $\bar{D}$ is a C^k manifold with boundary, and $\partial\bar{D}$ coincides with the topological boundary $\partial D = \bar{D} - D$ of D.

This is an immediate consequence of the rank theorem.

2.8.11 Remark. Note that if we orient R^n_+ by the usual n-form $dx_1 \wedge \ldots \wedge dx_n$ and R^{n-1} by the form $dx_2 \wedge \ldots \wedge dx_n$, then the orientation of ∂R^n_+ induced according to proposition 2.8.9 is the opposite of the usual orientation.

§ 2.9 Integration

Before we introduce integration on any orientable manifold, we must prove the formula for change of variables in a multiple integral. The proof that follows is due to J. Schwartz [1954].

2.9.1 Theorem. Let Ω, Ω' be open sets in R^n and $h: \Omega' \to \Omega$ a C^1 diffeomorphism. Then, if f is a continuous function with compact support in Ω, we have

2.9.2
$$\int_\Omega f(x)\mathrm{d}x_1 \ldots \mathrm{d}x_n = \int_{\Omega'} f \circ h(y)|\det \mathrm{d}h(y)|\mathrm{d}y_1 \ldots dy_n.$$

($|\det \mathrm{d}h(y)|$ is, of course, the absolute value of the determinant of the differential $\mathrm{d}h(y)$ at y).

Proof. We prove the result first when h is a linear transformation. Let A denote the matrix of h with respect to the canonical basis of R^n. By the theorem on elementary divisors (see e.g. Bourbaki [1952]. A can be written as a product of finitely many matrices A_ν each of which is either a diagonal matrix (with non-zero diagonal elements) or an elementary matrix, i.e., the matrix corresponding to one of the linear transformations

(a) $(x_1, \ldots, x_n) \mapsto$
$(x_1, \ldots, x_{j-1}, x_k, x_{j+1}, \ldots, x_{k-1}, x_j, x_{k+1}, \ldots, x_n)$

[i.e. interchange x_j and $x_k, j < k$],

(b) $(x_1, x_2, \ldots, x_n) \mapsto (x_1 + x_2, x_2, \ldots, x_n)$.

It is clearly sufficient to prove (2.9.2) for each of these transformations individually. For diagonal matrices and the transformation (a), this is a trivial consequence of Fubini's theorem. For the transformation (b), we have

$$\int_{\Omega'} f \circ h(y)|\det dh(y)|dy_1 \ldots dy_n$$

$$= \int_{R^n} f(x_1+x_2, x_2, \ldots, x_n)dx_1 \ldots dx_n$$

$$= \int_{R^{n-1}} dx_2 \ldots dx_n \int_R f(x_1+x_2, \ldots, x_n)dx_1$$

$$= \int_{R^{n-1}} dx_2 \ldots dx_n \int_R f(x_1, \ldots, x_n)dx_1$$

$$= \int_{R^n} f(x)dx_1 \ldots dx_n = \int_\Omega f(x)dx_1 \ldots dx_n.$$

Now, to prove the result in general, we proceed as follows. It is sufficient to prove that for any continuous *non-negative* f with compact support we have

2.9.3 $$\int_\Omega f(x)dx_1 \ldots dx_n \leqq \int_{\Omega'} (f \circ h)(y)|\det dh(y)|dy_1 \ldots dy_n$$

To see this, we remark that (2.9.3), applied, with Ω, Ω' interchanged, to the transformation h^{-1} and the function $f \circ h$ on Ω' gives us (2.9.2) for non-negative f. Since any continuous function with compact support is the difference of two non-negative continuous functions with compact support, this gives us (2.9.2) in general. Further, it suffices to prove the following statement: If Q is a closed cube (with equal sides) contained in Ω' we have

2.9.4 $$m(h(Q)) \leqq \int_Q |\det dh(y)|dy_1 \ldots dy_n,$$

where $m(S)$ is the Lebesgue measure of the measurable set S in R^n.

In fact, this implies (2.9.3) since the non-negative continuous function $f \circ h$ with compact support is the uniform limit of finite linear combinations $\sum c_j \chi_{Q_j}$, where $c_j \geqq 0$ and Q_j are closed cubes (with equal sides) and χ_S is the characteristic function of S (which is 1 on S and 0 outside).

PROOF OF (2.9.4). Let K be a closed cube in Ω' having equal sides δ. If $M = (a_{ij})$ is an $n \times n$ matrix, we set

$$\|M\| = \max_i \sum_j |a_{ij}|.$$

If I is the unit matrix, we have $\|I\| = 1$. If $l\colon R^n \to R^n$ is a linear map with matrix M, we write $\|l\| = \|M\|$.

The map $h\colon \Omega' \to \Omega$ is given by n functions, $h = (h_1, \ldots, h_n)$. For $x, y \in K$ we have, by theorem 1.1.9,

$$h_j(x) - h_j(y) = \sum_k \frac{\partial h_j}{\partial x_k}(t_j)(x_k - y_k),$$

where $t_j \in K$. Hence

$$|h_j(x) - h_j(y)| \leqq \delta\|(\mathrm{d}h)(t_j)\| \leqq \delta \sup_{a \in K} \|\mathrm{d}h(a)\|.$$

It follows at once that

2.9.5 $$m(h(K)) \leqq \delta^n\{\sup_{a \in K} \|\mathrm{d}h(a)\|\}^n = m(K) \sup_{a \in K} \|\mathrm{d}h(a)\|^n.$$

Let $a \in K$ and l be the linear transformation of R^n which is defined by

$$l^{-1} = (\mathrm{d}h)(a).$$

Let $g = \iota \circ h$. Then

$$m(g(K)) = |\det l|^{-1} m(h(K))$$

because of (2.9.2) for linear transformations. If we apply (2.9.5) to the transformation g, we obtain

2.9.6 $$m(h(K)) \leqq |\det \mathrm{d}h(a)| m(K) \sup_{y \in K} \|\mathrm{d}h(a)^{-1} \circ \mathrm{d}h(y)\|^n.$$

Clearly $\mathrm{d}h(a)^{-1} \circ \mathrm{d}h(y)$ converges to I as $\delta \to 0$, uniformly for y in a compact subset of Ω'. Hence there is a function $\lambda\colon R^+ \to R^+$ such that $\lambda(\delta) \to 0$ as $\delta \to 0$ and

$$\sup_{y \in K} \|\mathrm{d}h(a)^{-1} \circ \mathrm{d}h(y)\|^n \leqq 1 + \lambda(\delta), \qquad y \in Q$$

[Q a closed cube in Ω']. We divide Q into N^n cubes K_j of side $\delta =$

(side of $Q)/N$. Let $a_j \in K_j$. Then, by (2.9.6), we have

$$m(h(Q)) \leqq \sum_j m(h(K_j)) \leqq (1+\lambda(\delta)) \sum_j |\det dh(a_j)| m(K_j).$$

As $N \to \infty$, the term on the right converges to

$$\int_Q |\det dh(y)| dy_1 \ldots dy_n,$$

which proves (2.9.4) and with it the theorem.

We now pass to the consideration of integration on oriented manifolds. Let V be an orientable C^k manifold of dimension n ($k \geqq 1$) which may or may not have a boundary. Let ω be a continuous n-form on V with compact support. We suppose that V has a countable base. We suppose V oriented, so that there is given, on V, an n-form ω_0 without zeros. Let (U, φ), (U, ψ) be two coordinate systems (with the same U)). We write

$$\varphi(x) = (x_1, \ldots, x_n), \qquad \psi(x) = (y_1, \ldots, y_n)$$

and set

$$\theta_1 = dx_1 \wedge \ldots \wedge dx_n, \qquad \theta_2 = dy_1 \wedge \ldots \wedge dy_n.$$

Then

$$\theta_j = f_j \omega_0, \qquad f_j \in C^0(U), \qquad j = 1, 2.$$

We suppose that the coordinates are so chosen that $f_j > 0$. (Replace x_n or y_n by $-x_n$ or $-y_n$ if necessary). Such coordinate systems will be called *positive* relative to ω_0. Now, we have

$$\omega = g_1 \theta_1 = g_2 \theta_2, \qquad g_j \in C^0(U), \qquad j = 1, 2.$$

2.9.7 REMARK. We assert that if g_1 (and hence g_2) has compact support in U, then we have

$$\int_{\varphi(U)} g_1 \circ \varphi^{-1} dx_1 \ldots dx_n = \int_{\psi(U)} g_2 \circ \psi^{-1} dy_1 \ldots dy_n.$$

In fact, if we set

$$u(x) = \det d(\varphi \circ \psi^{-1})(\psi(x)), \qquad x \in U,$$

we have

$$\theta_1(x) = u(x) \theta_2(x)$$

and hence $u(x) = f_1(x)/f_2(x) > 0$. We can now apply theorem 2.9.1 and the fact that

$$g_1 \circ \varphi^{-1} \circ (\varphi \circ \psi^{-1})(\psi(x)) \mathrm{d}(\varphi \circ \psi^{-1})(\psi(x)) = g_2 \circ \psi^{-1}(\psi(x));$$

this gives us the assertion 2.9.7.

We can now define the integral of ω on V. Let $\{(U_i, \varphi_i)\}$ be a family of coordinate systems on V, $V = \bigcup U_i$, such that if θ_i is the form $\mathrm{d}x_1 \wedge \ldots \wedge \mathrm{d}x_n$ associated with $\varphi_i(x) = (x_1, \ldots, x_n)$, then $\theta_i = f_i \omega_0$ on U_i, where $f_i > 0$. Let $\{\eta_i\}$ be a partition of unity subordinate to $\{U_i\}$. Let

$$\omega = g_i \theta_i \quad \text{on} \quad U_i.$$

We define

2.9.8
$$\int_V \omega = \sum_i \int_{\varphi_i(U_i)} (\eta_i g_i) \circ \varphi_i^{-1} \mathrm{d}x_1 \ldots \mathrm{d}x_n.$$

It is an immediate consequence of remark 2.9.7 that this is independent of the coordinate systems $\{(U_i, \varphi_i)\}$ and the partition of unity $\{\eta_i\}$ chosen. Also if ω_0' is another form defining the same orientation as ω_0, the integral is unchanged.

2.9.9 STOKES' THEOREM. Let V be an oriented C^k manifold of dimension n with a countable base. Let ω be a C^1 $(n-1)$-form on V with compact support. Then we have

$$\int_{\partial V} \omega = \int_V \mathrm{d}\omega.$$

In particular, the above formula holds for all C^1 forms ω if V is compact.

PROOF. Let $\{(U_i, \varphi_i)\}$ be a family of coordinate systems such that

$$\theta_i = \mathrm{d}x_1 \wedge \ldots \wedge \mathrm{d}x_n = f_i \omega_0,$$
$$\varphi_i(x) = (x_1, \ldots, x_n),$$

where $f_i > 0$. Let $\{\eta_i\}$ be a partition of unity subordinate to $\{U_i\}$. We have only to prove that

$$\int_{\partial V} \eta_i \omega = \int_V \mathrm{d}(\eta_i \omega).$$

We consider two cases.

Case I. Suppose that $U_i \cap \partial V = \emptyset$; then $\varphi_i(U_i)$ is an open set in R^n and

$$\int_{\partial V} \eta_i \omega = 0.$$

We can suppose that U_i is an open set in R^n (by remark 2.9.7). Let

$$\eta_i \omega = \sum_{j=1}^{n} g_j \mathrm{d}x_1 \wedge \ldots \wedge \widehat{\mathrm{d}x_j} \wedge \ldots \wedge \mathrm{d}x_n,$$

where the $\wedge$ over the $\mathrm{d}x_j$ means that this term is to be omitted. We then have

$$\mathrm{d}(\eta_i \omega) = \sum_{j=1}^{n} \frac{\partial g_j}{\partial x_j} (-1)^{j-1} \mathrm{d}x_1 \wedge \ldots \wedge \mathrm{d}x_n,$$

so that

$$\int_V \mathrm{d}(\eta_i \omega) = \sum_{j=1}^{n} \int_{U_i} (-1)^{j-1} \frac{\partial g_j}{\partial x_j} \mathrm{d}x_1 \wedge \ldots \wedge \mathrm{d}x_n$$
$$= \sum_{j=1}^{n} (-1)^{j-1} \int_{R^n} \frac{\partial g_j}{\partial x_j} \mathrm{d}x_1 \ldots \mathrm{d}x_n = 0$$

since g_j has compact support for each j, so that

$$\int_R \frac{\partial g_j}{\partial x_j} \mathrm{d}x_j = 0.$$

Case II. $U_i \cap \partial V \neq \emptyset$. In this case $\varphi_i(U_i)$ is an open set in R^n_+ and

$$\varphi_i(U_i) \cap \{x \in R^n_+ | x_1 = 0\} \neq \emptyset.$$

We can again suppose that U_i is an open set in R^n_+. As in case I, if

$$\eta_i \omega = \sum_{j=1}^{n} g_j \mathrm{d}x_1 \wedge \ldots \wedge \widehat{\mathrm{d}x_j} \wedge \ldots \wedge \mathrm{d}x_n,$$
$$\mathrm{d}(\eta_i \omega) = \sum_{j=1}^{n} (-1)^{j-1} \frac{\partial g_j}{\partial x_j} \mathrm{d}x_1 \wedge \ldots \wedge \mathrm{d}x_n$$

we have

$$\int_{R^n_+} \frac{\partial g_j}{\partial x_j} \mathrm{d}x_1 \ldots \mathrm{d}x_n = 0 \qquad \text{if} \quad j \neq 1.$$

Moreover, if $j \neq 1$,

$$g_j \mathrm{d}x_1 \wedge \ldots \wedge \widehat{\mathrm{d}x_j} \wedge \ldots \wedge \mathrm{d}x_n | \partial V = 0$$

(see remark 2.5.16). Also

$$\int_{R^n{}_+} \frac{\partial g_1}{\partial x_1} \mathrm{d}x_1 \ldots \mathrm{d}x_n = \int_{-\infty}^{\infty} \mathrm{d}x_n \int_{-\infty}^{\infty} \mathrm{d}x_{n-1} \ldots \int_0^{\infty} \frac{\partial g_1}{\partial x_1} \mathrm{d}x_1$$

$$= -\int_{R^{n-1}} g_1(0, x_2, \ldots, x_n) \mathrm{d}x_2 \ldots \mathrm{d}x_n,$$

so that

$$\int_V \mathrm{d}(\eta_i \omega) = -\int_{R^{n-1}} g_1(0, x_2, \ldots, x_n) \mathrm{d}x_2 \ldots \mathrm{d}x_n = \int_{\partial V} \eta_i \omega$$

(see remark 2.8.11). This proves the result.

§ 2.10 One parameter groups

In this section V will denote a C^k manifold of dimension n with $k \geqq 3$ having a countable basis.

2.10.1 DEFINITION. Let $g: R \times V \to V$ be a C^r map. For $t \in R$, let $g_t: V \to V$ be the map $x \mapsto g(t, x)$. We call g a *one parameter group* of C^r transformations of V if g_t is a C^r diffeomorphism of V onto itself for each t and, for $t, s \in R$, we have

$$g_{s+t} = g_s \circ g_t;$$

in particular $g_0 =$ identity.
Let U be open in V, let $\varepsilon > 0$ and $I = \{t \in R |\ |t| < \varepsilon\}$. Let $g: I \times U \to V$ be a C^r map and $g_t: U \to V$ be, as above, the map $x \mapsto g(t, x)$. Then g is called a local one parameter group of C^r transformations of U into V if for each $t \in I$, g_t is a C^r diffeomorphism of U onto an open subset of V, g_0 is the identity, and, whenever s, t, $s+t \in I$ and x, $g_t(x) \in U$, we have

$$g_{s+t}(x) = g_s \circ g_t(x).$$

Let $g: I \times U \to V$ be a local one parameter group of C^r transformations. We define a C^{r-1} vector field $X = X_g$ on U as follows.

Let $a \in U$ and $f \in C_{a,k}$. We set

$$X(a)(f) = \left.\frac{\mathrm{d}(f \circ g_t(a))}{\mathrm{d}t}\right|_{t=0};$$

note that since g_0 is the identity, $g_t(a)$ is close to a for small enough t. Conversely, we have:

2.10.2 Proposition. Let X be a C^{k-1} vector field on V and $a \in V$. Then there is a neighbourhood U of a and a local one parameter group g of transformations of U into V such that X is induced by g, on U, i.e., $X = X_g$ on U.

Proof. It is sufficient to prove this proposition when V is an open set in R^n. Let the vector field X be given by

$$X(x) = \sum_{\nu=1}^{n} a_\nu(x) \left(\frac{\partial}{\partial x_\nu}\right)_x, \qquad x \in V.$$

By theorem 1.8.12, there is a $\delta > 0$ and a neighbourhood U_0 of a and a C^{k-1} map

$$g: I_0 \times U_0 \to V \qquad (I_0 = \{t \in R \mid |t| < \delta\}),$$

such that, if we set $a(x) = (a_1(x), \ldots, a_n(x))$ for $x \in V$, we have

$$\frac{\partial g}{\partial t}(t, x) = a(g(t, x)), \qquad g(0, x) = x, \qquad x \in U.$$

We claim that g is a local one parameter group. To prove this we set $g_t(x) = g(t, x)$, and choose

$$I = \{t \mid |t| < \varepsilon\}, \qquad \varepsilon > 0$$

and a neighbourhood U of a so small that

$$g_{s+t}(U) \subset U_0 \qquad \text{for} \quad s, t \in I.$$

For fixed $s \in I$, set $h_t = g_t \circ g_s$ on U. We see at once that h_t, g_{t+s} both satisfy the differential equation

$$\frac{\partial u(t, x)}{\partial t} = a(u(t, x)), \qquad u(0, x) = g_s(x), \qquad x \in U, t \in I.$$

Hence, by the uniqueness assertion in theorem 1.8.4, we have

$$h_t = g_{t+s}, \quad \text{i.e.,} \quad g_{t+s} = g_t \circ g_s.$$

In particular $g_t \circ g_{-t} = g_0 =$ identity, so that each g_t is a C^{k-1} diffeomorphism.

If $f \in C_{b,k}$, $b \in U$, we have

$$\left.\frac{\mathrm{d}(f \circ g_t(b))}{\mathrm{d}t}\right|_{t=0} = \sum \frac{\partial f}{\partial x_\nu}(b) \left.\frac{\mathrm{d}g_{t,\nu}(b)}{\mathrm{d}t}\right|_{t=0}$$

$$[g_t = (g_{t,1}, \ldots, g_{t,n})]$$

$$= \sum a_\nu(b) \frac{\partial f}{\partial x_\nu}(b) = X(b)(f),$$

so that the one parameter group g induces X on U.

2.10.3 REMARK. The local one parameter group g is unique in an obvious sense, as follows from the uniqueness assertion in theorem 1.8.4.

2.10.4 THEOREM. *Let X be a C^{k-1} vector field with compact support on the V. Then there is a unique one parameter group g of C^{k-1} transformations of V which induces X on V; further $g(t, x) = x$ for all t if x is outside a compact subset of V.*

PROOF. Let K be a compact set such that $X(a) = 0$ for $a \notin K$. By proposition 2.10.2, for any $a \in K$, there is a neighbourhood U_a and a local one parameter group $g_t^{(a)}$: $U_a \to V$, $|t| < \varepsilon(a)$, which induces X on U_a. Choose a_ν, with $1 \leqq \nu \leqq p$ such that

$$U = \bigcup_{1 \leqq \nu \leqq p} U_{a_\nu} \supset K$$

and let

$$\varepsilon = \min_\nu \varepsilon(a_\nu).$$

If $U_{a_\nu} \cap U_{a_\mu} \neq \emptyset$, then $g_t^{(a_\nu)}$, $g_t^{(a_\mu)}$ induce on $U_{a_\nu} \cap U_{a_\mu}$ the same vector field, hence coincide for $|t| < \varepsilon$. Hence we can define g_t on U by $g_t(x) = g_t^{(a_\nu)}(x)$ if $x \in U_{a_\nu}$. Further, if $x \in U$, $x \notin K$, then $X(x) = 0$, so that for all $f \in C_{x,k}$ we have

$$\frac{\mathrm{d}f \circ g_t(x)}{\mathrm{d}t} = 0, \qquad \text{for} \quad t = 0.$$

It follows from the uniqueness assertion in theorem 1.8.4 that $g_t(x) \equiv x$. Hence, we can extend $g: I \times U \to V$ to a map $g: I \times V \to V$ by setting $g(t, x) = x$ for $x \notin K$; here $I = \{t \mid |t| < \varepsilon\}$. Moreover, for any $x \in V$, if $t, s, t+s \in I$, we have $g_{t+s}(x) = g_t \circ g_s(x)$.

If now $t \in \mathsf{R}$ is arbitrary, we choose an integer $p > 0$ such that $t' = t/p \in I$ and define

$$g_t = g_{t'} \circ \ldots \circ g_{t'},$$

($g_{t'}$ composed with itself p times).
It is immediately seen that this defines a C^{k-1} map $g: \mathsf{R} \times V \to V$ which is a one-parameter group and extends our map $g: I \times V \to V$; in particular, g induces the vector field X.

2.10.5 Remark. Let U be an open subset of V and $\sigma: U \to V$ a C^r diffeomorphism of V onto an open subset of V. Let X be a vector field on U. Then σ induces a vector field $\sigma_*(X)$ on $U' = \sigma(U)$, where

$$\sigma_*(X)(\sigma(a)) = \sigma_{*,a}(X(a))$$

(see remark 2.4.5). If $f \in C^k(U')$, we have

$$\sigma_*(X)(f) = X(f \circ \sigma) \circ \sigma^{-1}.$$

Hence, if X, Y are two vector fields on U, and $f \in C^k(U')$, we have

$$\begin{aligned}[\sigma_*(X), \sigma_*(Y)](f) &= \sigma_*(X)\{Y(f \circ \sigma) \circ \sigma^{-1}\} - \sigma_*(Y)\{X(f \circ \sigma) \circ \sigma^{-1}\} \\ &= (X(Y(f \circ \sigma)) - Y(X(f \circ \sigma))) \circ \sigma^{-1} = \sigma_*([X, Y])(f),\end{aligned}$$

so that we have

2.10.6 $$\sigma_*([X, Y]) = [\sigma_*(X), \sigma_*(Y)].$$

Let now $\sigma: U \to U'$ be a C^r diffeomorphism as above. Let $W \Subset U$ and $W' = \sigma(W)$. Let $g: I \times U \to V$ be a local one parameter group, which induces the vector field X on U. Then, if I is small enough, $g(I\times W)\subset U'$, so that $t\mapsto \sigma\circ g_t\circ\sigma^{-1}$ defines a local one parameter group $g'\colon I\times W'\to V$. We have, if $f\in C^k(W')$ and $a\in W'$

$$\sigma_*(X)(f)(a) = X(f\circ\sigma)\circ\sigma^{-1}(a) = \left.\frac{\mathrm{d}(f\circ\sigma\circ g_t\circ\sigma^{-1}(a))}{\mathrm{d}t}\right|_{t=0}$$

so that g' induces, on W', the vector field $\sigma_*(X)$. From this, we conclude:

2.10.7 COROLLARY. If $\sigma(W) = W'$ is also relatively compact in U, then, we have

$$\sigma\circ g_t(x) = g_t\circ\sigma(x)$$

for $x\in W$ and all small t if and only if for $a\in W$, we have

$$\sigma_{*,a}(X(a)) = X(\sigma(a)).$$

2.10.8 DEFINITION. Let $g\colon I\times U\to V$ be a local one parameter group of C^r transformations, $r > 2$ and X a vector field on U. We say that *g leaves X invariant* if for any $a\in U$, we have

$$(g_t)_{*,a}(X(a)) = X(g_t(a))$$

for all small enough t.

Let now $g\colon I\times U\to V$ be a local one parameter group of C^r transformations as above, and let U' be an open set $U'\Subset U$. Let Y be a C^r vector field on U, $r\geqq 2$. For all small enough t, we define a vector field Y_t on U' by

$$Y_t(f) = Y(f\circ g_t)\circ g_{-t} = (g_t)_*(Y)(f);$$

and a vector field $\mathrm{d}Y_t/\mathrm{d}t$ by $(\mathrm{d}Y_t/\mathrm{d}t)(f) = \mathrm{d}Y_t(f)/\mathrm{d}t$.

2.10.9 PROPOSITION. We assert that, for small enough t, we have on U'

$$\mathrm{d}Y_t/\mathrm{d}t = [Y_t, X],$$

where X is the vector field on U induced by g.

PROOF. We set $Z_t = \mathrm{d}Y_t/\mathrm{d}t$. If $f \in C^k(U)$, we have

$$
\begin{aligned}
Z_0(f) &= \lim_{t\to 0} t^{-1}\{Y(f\circ g_t)\circ g_{-t} - Y(f)\} \\
&= \lim_{t\to 0} t^{-1}\{Y(f\circ g_t) - Y(f) - Y(f)\circ g_t + Y(f)\}\circ g_{-t} \\
&= \lim_{t\to 0} t^{-1} Y(f\circ g_t - f) - \lim_{t\to 0} t^{-1}(Y(f)\circ g_t - Y(f))
\end{aligned}
$$

if these last two limits exist uniformly on U', since $\lim_{t\to 0} g_{-t}$ is the identity. By definition of X, we have

$$X(Y(f)) = \lim_{t\to 0} t^{-1}\{Y(f)\circ g_t - Y(f)\},$$

uniformly on U'. Let now $h(t, x) = f\circ g_t(x)$. Clearly $h \in C^2(I\times U')$ if I is a small enough interval about 0 in $\mathbf{R}$. Hence the function

$$
F = \begin{cases} t^{-1}(f\circ g_t - f) = t^{-1}(h(t,x) - h(0,x)) & \text{for } t \neq 0 \\ \dfrac{\partial h}{\partial t}(0, x) & \text{for } t = 0 \end{cases}
$$

is in $C^1(I\times U')$. Hence

$$\lim_{t\to 0} t^{-1} Y(f\circ g_t - f) = Y\big(\lim_{t\to 0} t^{-1}(f\circ g_t - f)\big) = Y\big(X(f)\big).$$

Hence for $f\in C^k(U)$:

$$Z_0(f) = [Y, X](f) = [Y_0, X](f).$$

Clearly, this implies that $Z_0 = [Y_0, X]$ on U'.

If now t_0 is small enough, we have

$$
\begin{aligned}
(g_{t_0})_* Z_0 = Z_t \quad \text{and} \quad (g_{t_0})_*[Y_0, X] &= [(g_{t_0})_* Y_0, (g_{t_0})_* X] \\
&= [Y_{t_0}, X] \quad \text{on} \quad U',
\end{aligned}
$$

which proves the proposition.

2.10.10 PROPOSITION. Let $g, h: I\times U \to V$ be local one parameter groups which induce C^r vector fields X, Y on $U (r \geqq 2)$. Then, for any $U' \Subset U$, we have $g_t\circ h_s(x) = h_s\circ g_t(x)$ for $x\in U'$ and small enough t, s if and only if $[X, Y] = 0$ on U.

PROOF. If g_t, h_s commute on U' for small t, s, we see at once that

g_t leaves Y invariant (definition 2.10.8). Hence

$$0 = \left.\frac{\mathrm{d}Y_t}{\mathrm{d}t}\right|_{t=0} = [Y, X] \quad \text{on} \quad U'$$

(by proposition 2.10.9). Since $U' \Subset U$ is arbitrary,

$$[X, Y] = -[Y, X] = 0.$$

Conversely if $[X, Y] = 0$, we have

$$\mathrm{d}Y_t/\mathrm{d}t = (g_t)_*[Y, X] = 0$$

(by proposition 2.10.9) and g_t leaves Y invariant. The result then follows from corollary 2.10.7.

2.10.11 Remark. All the results of this section have analogues for a complex analytic manifold V and holomorphic vector fields. One introduces holomorphic (local) one parameter groups, which are *holomorphic* maps g: $(I \times U) \subset \mathbf{C} \times V \to V$ with the obvious properties. Then holomorphic vector fields and holomorphic local one parameter groups correspond to one another as before. Commutation of the elements of the groups are again expressed in terms of the vanishing of brackets.

The proofs are identical with the ones given above, and are therefore omitted.

For all this material, see Nomizu [1956].

§ 2.11 The Frobenius theorem

Let V be a C^k manifold of dimension n with a countable basis ($k \geqq 3$).

2.11.1 Definition. A *differential system* or *distribution* $\mathfrak{D}$ of rank p on V is an assignment to each point $a \in V$ of a subspace $\mathfrak{D}(a) \subset T_a(V)$ of dimension p. $\mathfrak{D}$ is called *differentiable of class* C^r ($0 \leqq r < k$) if every $a \in V$ has a neighbourhood U in which there are C^r vector fields $X_1, \ldots, X_p$ such that $X_1(b), \ldots, X_p(b)$ form a basis of $\mathfrak{D}(b)$ for $b \in U$. The X_j are said to *generate* $\mathfrak{D}$ on U.

2.11.2 Definition. Let $\mathfrak{D}$ be a differential system of rank p. A submanifold i: $W \to V$ is called an *integral* (or integral manifold) of $\mathfrak{D}$

if for any $a \in W$, we have

$$i_{*,a}(T_a(W)) \subset \mathfrak{D}(i(a)).$$

We shall also say that a C^r map $f\colon V' \to V$ is an integral of $\mathfrak{D}$ if

$$f_{*,a}(T_a(V')) \subset \mathfrak{D}(f(a))$$

for $a \in V'$. Note that a submanifold of an integral is again an integral.

2.11.3 Definition. We say that $\mathfrak{D}$ is completely integrable if for any $a \in V$, there is a coordinate system (U, φ), $a \in U$, $\varphi(x) = (x_1, \ldots, x_n)$ such that for all $c_j, p < j \leqq n$, the submanifolds given by

$$U_c = \{x \in U | x_j = c_j, p < j \leqq n\}$$

are integrals of $\mathfrak{D}$.

Let $\mathfrak{D}$ be a completely integrable system, and let $a \in V$. Choose a coordinate system (U, φ), $a \in U$, such that definition 2.11.3 holds. Then we have:

2.11.4 Proposition. If $i\colon\ W \to U$ is an integral and W is connected, then $i(W) \subset U_c$ for some $c = (c_{p+1}, \ldots, c_n)$.

Proof. Let $\varphi(x) = (x_1, \ldots, x_n)$. For any $b \in U$, it is clear that $T_b(U_c)$ $[b = (b_1, \ldots, b_p, c_{p+1}, \ldots, c_n)]$ has dimension p and $T_b(U_c) \subset \mathfrak{D}(b)$, so that $T_b(U_c) = \mathfrak{D}(b)$. Further $T_b(U_c)$ consists of those vectors in $T_b(V)$ which are annihilated by $(\mathrm{d}x_j)_b$, $p < j \leqq n$. Hence, if $i\colon W \to U$ is an integral, then $i^*(\mathrm{d}x_j) = 0$, $p < j \leqq n$. Since W is connected, this implies that $x_j \circ i$ is constant on W, which proves our assertion.

In particular, *the submanifolds* $\{U_c\}$, *c sufficiently small, are independent of the coordinate system* (U, φ).

2.11.5 Definition. Let $\mathfrak{D}$ be a C^r differential system. We say that $\mathfrak{D}$ is *involutive* if for any $a \in V$, there is a neighbourhood U and C^r vector fields $X_1, \ldots, X_p$ generating $\mathfrak{D}$ on U such that for any $b \in U$ we have

$$[X_\mu, X_\nu](b) \in \mathfrak{D}(b) \qquad \text{for} \quad 1 \leqq \mu, \nu \leqq p.$$

2.11.6 REMARK. Note that this is equivalent to the following: For any open set $U \subset V$ and two C^r vector fields X, Y on U such that $X(a)$, $Y(a) \in \mathfrak{D}(a)$ for $a \in U$, we have

$$[X, Y](a) \in \mathfrak{D}(a) \qquad \text{for} \quad a \in U.$$

2.11.7 PROPOSITION. If $\mathfrak{D}$ is an involutive C^r differential system of rank p, then any $a \in V$ has a neighbourhood U in which there are vector fields X_ν, $1 \leqq \nu \leqq p$ such that the $X_\nu(b)$ generate $\mathfrak{D}(b)$ for $b \in U$ and $[X_\nu, X_\mu] = 0$ in U.

PROOF. Let $a \in V$ and (U, φ) be a coordinate system with $a \in V$. Let $\varphi(x) = (x_1, \ldots, x_n)$. If U is small, then there are C^r vector fields $Y_1, \ldots, Y_p$ such that $Y_\nu(x)$ generate $\mathfrak{D}(x)$ for $x \in U$. Let

$$Y_\nu(x) = \sum_{\mu=1}^{n} a_{\nu\mu}(x) \left(\frac{\partial}{\partial x_\mu}\right)_x; \qquad a_{\nu\mu} \in C^r(U).$$

Since the $\{Y_\nu(a)\}$ generate a vector space of rank p, the matrix $(a_{\nu\mu}(a))$ for $1 \leqq \nu \leqq p$ and $1 \leqq \mu \leqq n$ has rank p. We may assume, without loss of generality, that if

$$A(x) = (a_{\nu\mu}(x)), \qquad 1 \leqq \nu \leqq p, 1 \leqq \mu \leqq p,$$

then $A(a)$ has rank p. If U is small enough, the matrix $A(x)$ is invertible for $x \in U$. Let $B(x) = (b_{\nu\mu}(x)) = A(x)^{-1}$; then $b_{\nu\mu} \in C^r(U)$ for $1 \leqq \nu, \mu \leqq p$. Let

$$X_\nu = \sum_{\mu=1}^{p} b_{\nu\mu} Y_\mu .$$

Then X_ν has the form

$$X_\nu = \frac{\partial}{\partial x_\nu} + \sum_{\mu > p} c_{\nu\mu} \frac{\partial}{\partial x_\mu}, \qquad c_{\nu\mu} \in C^r(U);$$

moreover the $\{X_\nu\}$ form a basis of $\mathfrak{D}$ in U. Since $\mathfrak{D}$ is involutive, we have

$$[X_\nu, X_\mu] = \sum_{m=1}^{p} \lambda_m X_m, \qquad \lambda_m \in C^r(U).$$

Since

$$\left[\frac{\partial}{\partial x_\nu}, \frac{\partial}{\partial x_\mu}\right] = 0,$$

we see at once that if

$$[X_\nu, X_\mu] = \sum_{m=1}^{n} \xi_m \frac{\partial}{\partial x_m},$$

then $\xi_m = 0$ for $1 \leqq m \leqq p$. Trivially $\lambda_m = \xi_m$ for $m \leqq p$, hence $= 0$. Thus $[X_\nu, X_\mu] = 0$.

2.11.8 THEOREM. Let $X_1, \ldots, X_p$ be C^r vector fields, $r \geqq 2$, on V which are linearly independent at every point of V and such that $[X_\nu, X_\mu] = 0$. Then for any $a \in V$ there exists a C^r coordinate system (U, φ), $a \in U$, such that if $\varphi(x) = (x_1, \ldots x_p)$ and $\partial/\partial x_1, \ldots, \partial/\partial x_n$ are the associated vector fields in U, we have $X_\nu = \partial/\partial x_\nu$, $\nu = 1, \ldots, p$.

PROOF. Let (U', φ') be a coordinate system at a such that the vectors

$$X_1(a), \ldots, X_p(a), \left(\frac{\partial}{\partial x'_{p+1}}\right)_a, \ldots, \left(\frac{\partial}{\partial x'_n}\right)_a$$

are linearly independent; here $\varphi'(x) = (x'_1, \ldots, x'_n)$, and $\partial/\partial x'_j$ denotes the associated vector fields in U'. [It is clear that such a system exists; one has at most to subject R^n to a linear transformation.] We suppose that $\varphi'(a) = 0$.
Let

$$g^{(\nu)}: I \times U' \to V$$

be local one parameter groups of C^r transformations inducing X_ν on U'; here $I = \{t \in R \mid |t| < \varepsilon\}$; The $g^{(\nu)}$ are uniquely determined if U' is small enough (proposition 2.10.2 and remark 2.10.3).
Let $t_1, \ldots, t_p, x'_{p+1}, \ldots, x'_n$ be real numbers of absolute value $< \delta$, where $\delta > 0$. We define a map

$$h: \Omega \to V, \qquad \Omega = \{x \in R^n \mid |x| < \delta\}$$

by

$$h(t_1, \ldots, t_p, x'_{p+1}, \ldots, x'_n) =$$
$$g^{(1)}_{t_1} \circ \ldots \circ g^{(p)}_{t_p} \circ \varphi'^{-1}(0, \ldots, 0, x'_{p+1}, \ldots, x'_n).$$

This is well defined if δ is small enough. By definition, if $f \in C_{a,k}$,

$$\frac{\partial}{\partial t_1}(f \circ h)(0) = X_1(a)(f)$$

(since $g^{(1)}$ induces X_1). Since, by hypothesis $[X_\nu, X_\mu] = 0$, the $g^{(\nu)}_{t_\nu}$ 'commute' by proposition 2.10.10, so that we have

$$\frac{\partial}{\partial t_\nu}(f \circ h)(0) = X_\nu(a)(f), \qquad \nu = 1, \ldots, p,$$

i.e.,

$$h_{*,a}\left(\left(\frac{\partial}{\partial t_\nu}\right)_0\right) = X_\nu(a).$$

Moreover, with obvious notation,

$$h_{*,0}\left(\left(\frac{\partial}{\partial x'_j}\right)_0\right) = \left(\frac{\partial}{\partial x'_j}\right)_a \qquad \text{for} \quad p < j \leqq n.$$

This shows, in particular, that $h_{*,0}$ has rank n. Hence, by the inverse function theorem 2.2.10, if δ is small enough, h is a C^r isomorphism of Ω onto an open set U in V. Again by definition,

$$h_{*,u}\left(\left(\frac{\partial}{\partial t_1}\right)_u\right) = X_1(h(u)),$$

and, by the commutativity of the $g^{(\nu)}_{t_\nu}$, we have

$$h_{*,u}\left(\left(\frac{\partial}{\partial t_\nu}\right)_u\right) = X_\nu(h(u)), \qquad 1 \leqq \nu \leqq p.$$

The C^r coordinate system (U, h^{-1}) has the required properties.

2.11.9 THEOREM OF FROBENIUS: FIRST FORM. A C^r differential system $\mathfrak{D}$ on V with $r \geqq 2$ is completely integrable if and only if it is involutive.

PROOF. An involutive system is completely integrable by proposition 2.11.7 and theorem 2.11.8. Conversely, if $\mathfrak{D}$ is completely integrable and (U, φ) is a C^r coordinate system such that $\varphi(x) = (x_1, \ldots, x_n)$ and the sets

$$U_c = \{x \in U \mid x_j = c_j,\, p < j \leqq n\}$$

are integrals, then $(\partial/\partial x_\nu)_a$, $1 \leqq \nu \leqq p$, $a \in U_c$ span the tangent space $T_a(U_c)$, hence span $\mathfrak{D}(a)$ for $a \in U$; clearly, then, $\mathfrak{D}$ is involutive. [Here we have, however, only a C^{r-1} basis of $\mathfrak{D}$.]

2.11.10 REMARK. We have used essentially the fact that the vector fields under consideration are C^r, $r \geqq 2$ [in the proof of proposition 2.10.9]. However, also involutive C^1 systems are C^1 completely integrable. This can be proved using methods similar to the ones used above. One needs then the theorems of § 1.8, and the fact that in a system of equations $dx/dt = f(x, t)$, the solution x has one more derivative in t than in *all* the variables.

We now consider another method of defining differential systems. Let $\omega_{p+1}, \ldots, \omega_n$ be 1-forms on V which are linearly independent at every point. We define a differential system $\mathfrak{D}$ by setting

$$\mathfrak{D}(a) = \{X \in T_a(V) | \omega_\nu(a)(X) = 0 \quad \text{for} \quad p < \nu \leqq n\}.$$

If the ω_i are differentiable, then so is $\mathfrak{D}$. In fact, in terms of a suitable coordinate system, we may suppose that the ω_ν have the form

$$\omega_\nu = dx_\nu + \sum_{\mu \leqq p} a_{\nu\mu} dx_\mu, \qquad \nu > p, \quad a_{\nu\mu} \in C^r(U).$$

Then $\mathfrak{D}$ is the differential system spanned by the vector fields

$$X_\mu = \frac{\partial}{\partial x_\mu} - \sum_{\nu > p} a_{\nu\mu} \frac{\partial}{\partial x_\nu}, \qquad 1 \leqq \mu \leqq p,$$

since clearly $\omega_\nu(X_\nu) = 0$ and the X_μ are linearly independent. Moreover, locally, any differential system is obtained in this way.

2.11.11 THEOREM OF FROBENIUS: SECOND FORM. Let $\omega_{p+1}, \ldots, \omega_n$ be C^r 1-forms which are linearly independent at every point and $\mathfrak{D}$ the differential system defined by them. Then $\mathfrak{D}$ is completely integrable if and only if the following condition is satisfied:

Every $a \in V$ has a neighbourhood U in which there exist 1-forms $\alpha_{\mu\nu}$ such that

2.11.12
$$d\omega_\nu = \sum_{\mu=p+1}^{n} \omega_\mu \wedge \alpha_{\mu\nu};$$

i.e., $d\omega_\nu$ belongs to the ideal generated by the ω_μ.

PROOF. We remark that the condition (2.11.12) is invariant under change of basis. In other words, if $\{\omega'_\nu\}$ is another set of $n-p$ 1-forms generating the same system $\mathfrak{D}$, then there are C^r functions $a_{\mu\nu}$ such that

$$\omega'_\nu = \sum_\mu a_{\mu\nu}\,\omega_\nu .$$

It follows that if (2.11.12) holds, then $d\omega'_\nu$ belongs to the ideal generated by the ω'_μ.

If $\mathfrak{D}$ is completely integrable, and (U, φ) is a coordinate system as in definition 2.11.3, the tangent space $T_a(U_c)$, $a \in U_c$ is the space orthogonal to $(dx_{p+1})_a, \ldots, (dx_n)_a$. Hence $\mathfrak{D}|U$ is defined by the 1-forms $dx_{p+1}, \ldots, dx_n$. These forms are closed, so that the condition (2.11.12) for them is trivial.

Suppose conversely that (2.11.12) holds. Let $X_1, \ldots, X_p$ be vector fields in a neighbourhood of a generating $\mathfrak{D}$. Then, by proposition 2.6.6 we have

$$(d\omega_\nu)(X_\kappa, X_\mu) = X_\kappa\,\omega_\nu(X_\mu) - X_\mu\,\omega_\nu(X_\kappa) - \omega_\nu([X_\kappa, X_\mu]).$$

Now, by hypothesis,

$$\omega_\nu(X_\mu) = \omega_\nu(X_\kappa) = 0$$

and by (2.11.12),

$$(d\omega_\nu)(X_\kappa, X_\mu) = 0.$$

Hence

$$\omega_\nu([X_\kappa, X_\mu]) = 0 \qquad \text{for} \quad \nu = p+1, \ldots, n,$$

so that $[X_\kappa, X_\mu](b) \in \mathfrak{D}(b)$ for $b \in U$. Hence $\mathfrak{D}$ is involutive and thus, by theorem 2.11.9, completely integrable.

Our next theorem asserts the existence of maximal integrals of completely integrable differential systems.

2.11.13 THEOREM. Let $\mathfrak{D}$ be a completely integrable C^r differential system of rank p on V. Then for any $a \in V$, there is a connected integral C^r submanifold (W, i) of V of dimension p with $a \in i(W)$ such that for any connected integral C^r submanifold $j\colon\ W' \to V$ of V with $a \in$

$j(W')$ there is a C^r map η: $W' \to W$ making W' a C^r submanifold of W such that $j = i \circ \eta$.

PROOF. Let I be the closed unit interval in R. A chain of C^r mappings $\gamma_\nu: I \to V$, $0 \leqq \nu \leqq N$ with $\gamma_0(0) = a$, $\gamma_N(1) = x$, $\gamma_{\nu+1}(0) = \gamma_\nu(1)$, $0 \leqq \nu < N$, is called an integral chain from a to x, if each γ_ν is an integral of $\mathfrak{D}$ (definition 2.11.2). Let W be the set of $x \in V$ such that there is an integral chain from a to x. Let $x_0 \in W$, and let (U, φ) be a coordinate system with $x_0 \in U$, such that if $\varphi(x) = (x_1, \ldots, x_n)$, the sets

$$U_c = \{x \in U | x_j = c_j, p < j \leqq n\}$$

are integrals of $\mathfrak{D}$, we suppose that $\varphi(U)$ is a cube, so that the U_c are connected, and that $\varphi(x_0) = 0$. Then, clearly $U_0 \subset W$. We topologize W by requiring that the sets U_0 so obtained form a fundamental system of neighbourhoods of x_0 in W. Clearly W is Hausdorff, and the injection i: $W \to V$ is continuous. If again U, φ, U_0 are as above, then if π_1 denotes the projection of R^n onto R^p, i.e., $\pi_i(x_1, \ldots, x_n) = (x_1, \ldots, x_p)$, then $\varphi_0 = \pi_1 \circ \varphi | U_0$ is a homeomorphism of U_0 onto an open set in R^p. The pairs (U_0, φ_0) define on W the structure of a C^r manifold, and the injection i: $W \to V$ makes of W a submanifold of V. Clearly i: $W \to V$ is an *integral* submanifold of $\mathfrak{D}$.

Let now $j : W' \to V$ be any connected integral C^r submanifold and $a' \in W'$ be such that $j(a') = a$. Let $w' \in W'$ and $\gamma'_0, \ldots, \gamma'_N$ a chain of C^r maps $\gamma'_\nu: I \to W'$ with $\gamma'_0(0) = a'$, $\gamma'_N(1) = w'$, $\gamma'_{\nu+1}(0) = \gamma'_\nu(1)$, with $0 \leqq \nu < N$. Let $\gamma_\nu = j \circ \gamma'_\nu$. Then the γ_ν form an integral chain from a to $j(w')$, so that $j(w') \in W$. We set $\eta(w') = j(w')$. This defines a map η: $W' \to W$. Clearly $i \circ \eta = j$. It follows at once from proposition 2.11.4 that η is continous. Hence, by proposition 2.5.12, η is a C^r map. Since, for $w' \in W'$, we have

$$j_{*, w'} = i_{*, \eta(w')} \circ \eta_{*, w'},$$

and $j_{*, w'}$ is injective, so is $\eta_{*, w'}$, so that η makes of W' a C^r submanifold of W.

We now give a third form of the Frobenius theorem in which it appears as a direct generalization of the existence theorem for ordinary differential equations, (§ 1.8).

2.11.14 THEOREM OF FROBENIUS: THIRD FORM. Let Ω be an open set in R^n, Ω' an open set in R^m. We denote a point of R^n by $x = (x_1, \ldots, x_n)$, a point of R^m by $t = (t_1, \ldots, t_m)$. Let $f_\nu\colon \Omega\times\Omega' \to R^n$ be C^k mappings, $k \geqq 2$, $\nu = 1, \ldots, m$. In order that to every $t_0 \in \Omega'$ and $x_0 \in \Omega$ these exist a neighbourhood U of t_0 and a unique C^k map $x\colon U \to \Omega$ such that

2.11.15 $$x(t_0) = x_0, \quad \frac{\partial x(t)}{\partial t_\nu} = f_\nu(x(t), t), \quad t \in U, \quad \nu = 1, \ldots, m,$$

it is necessary and sufficient that

2.11.16 $$\frac{\partial f_\nu}{\partial t_\mu}(x, t) + (\mathrm{d}_1 f_\nu)(x, t) f_\mu(x, t) = \frac{\partial f_\mu}{\partial t_\nu}(x, t) + (\mathrm{d}_1 f_\mu)(x, t) f_\nu(x, t)$$

for $1 \leqq \mu, \nu \leqq m$, $(x, t) \in \Omega\times\Omega'$. [$(\mathrm{d}_1 f_\nu)(a, b)$ is the linear map of R^n into itself defined by $(\mathrm{d}_1 f_\nu)(a, b) = (\mathrm{d}g)(a)$, where $g\colon \Omega \to R^n$ is the map $x \mapsto f_\nu(x, b)$; see definition 1.3.4.]

PROOF. The uniqueness of the solution, if it exists, follows from the corresponding uniqueness assertion for ordinary differential equations (theorem 1.8.4).

If (2.11.15) always has a solution, then (2.11.16) holds since both sides of the equation, at the point (x_0, t_0) are then equal to

$$\left.\frac{\partial^2 x(t)}{\partial t_\mu \partial t_\nu}\right|_{t=t_0}$$

To prove the converse, we proceed as follows. (2.11.15) can be written

2.11.17 $$\frac{\partial x_\kappa}{\partial t_\nu} = f_{\nu\kappa}(x, t), \quad x = (x_1, \ldots, x_n), \quad f_\nu = (f_{\nu 1}, \ldots, f_{\nu n}),$$
$$\kappa = 1, \ldots, n, \qquad \nu = 1, \ldots, m.$$

Consider the differential forms

2.11.18 $$\mathrm{d}x_\kappa - \sum_{\nu=1}^{m} f_{\nu\kappa}(x, t)\mathrm{d}t_\nu, \qquad \kappa = 1, \ldots, n$$

on $\Omega \times \Omega'$, and let $\mathfrak{D}$ be the differential system of rank m defined by them. It is obvious that if $\mathfrak{D}$ has an integral manifold of the form

$$x-\xi(t) = 0, \qquad \xi(t_0) = x_0,$$

where ξ is a C^k map of a neighbourhood of t_0 in Ω, then $x = \xi$ is a solution of (2.11.15).

If now $\mathfrak{D}$ is completely integrable, then there is a C^r submanifold of $\Omega \times \Omega'$ of dimension m in a neighbourhood of (x_0, t_0), say W, which is an integral of $\mathfrak{D}$. We can then find C^r functions $u_1, \ldots, u_n$ in a neighbourhood of (x_0, t_0) such that, near (x_0, t_0),

$$W = \{(x, t) | u_1(x, t) = \ldots = u_n(x, t) = 0\},$$

$du_\mu(x_0, t_0)$, $(\mu = 1, \ldots, n)$ being linearly independent. Clearly then $du_\mu(x_0, t_0)$, $\mu = 1, \ldots, n$ generate the same subspace of $T^*_{(x_0, t_0)}(\Omega \times \Omega')$ as the forms (2.11.18). This implies that the covectors $(d_1 u_\mu)(x_0, t_0)$, $\mu = 1, \ldots, n$ are linearly independent. It then follows, from the implicit function theorem 1.3.5 and corollary 1.3.9 that, in the neighbourhood of (x_0, t_0), W is defined by equations of the form $x-\xi(t)=0$; as already remarked, this implies that the equations (2.11.15) are solvable if $\mathfrak{D}$ is completely integrable.

Now, we have seen that $\mathfrak{D}$ is generated by the vector fields

$$X_\nu = \frac{\partial}{\partial t_\nu} + \sum_{\kappa=1}^{n} f_{\nu\kappa}(x, t) \frac{\partial}{\partial x_\kappa}, \qquad \nu = 1, \ldots, m;$$

see considerations preceding theorem 2.11.11. Hence, as in the proof of proposition 2.11.7, $\mathfrak{D}$ is completely integrable if and only if

$$[X_\nu, X_\mu] = 0, \qquad \nu, \mu = 1, \ldots, m.$$

But these are precisely the conditions (2.11.16). This proves the theorem.

2.11.19 Remark. The theorem is true also when the f_ν are C^1 mappings. This can be proved using remark 2.11.10.

If in the above theorem, we take $n = 1$ and for the f_ν functions independent of x, we obtain the following.

In order that there exists a C^k function $x(t_1, \ldots, t_m)$ in a neighbourhood of $t = t_0$ such that

$$\frac{\partial x}{\partial t_\nu} = f_\nu(t), \qquad \nu = 1, \ldots, m$$

it is necessary and sufficient that $\partial f_\nu / \partial t_\mu = \partial f_\mu / \partial t_\nu$. This can be formulated as follows: Let

$$\omega = \sum_{\nu=1}^{m} f_\nu(t) \mathrm{d}t_\nu,$$

then, in the neighbourhood of any point of Ω', there is a function f with $\mathrm{d}f = \omega$ if and only if $\mathrm{d}\omega = 0$.

This is a special case of Poincaré's lemma to be proved in § 2.13.

A different treatment of the Frobenius theorem in the form of theorem 2.11.9 will be found in CHEVALLEY [1944].

2.11.20 REMARK. All the results of this section have analogues for holomorphic vector fields, etc. on complex manifolds. One defines holomorphic differential systems $\mathfrak{D}$ by the condition that they are locally generated by holomorphic vector fields. Completely integrable systems are defined as in definition 2.11.3 with the help of complex coordinates. The theorems 2.11.8, 9, 11, 13, 14 all have analogues for holomorphic systems. Naturally, in (2.11.12), the $\alpha_{\mu\nu}$ are required to be *holomorphic* 1-forms.

§ 2.12 Almost complex manifolds

We have seen (remark 2.4.6) that the tangent space $T_a(V)$ to a complex manifold V at a point $a \in V$ carries a natural structure of a complex vector space. It is sometimes useful to consider *only* this structure. This leads to a more general class of manifolds.

Let V be a C^∞ manifold of dimension $n = 2m$ (the dimension is even). We suppose given, for each $a \in V$, the structure of a C vector space (of dimension m) on $T_a(V)$. We say that this structure depends differentiably on a if the following condition is satisfied: Any $a \in V$ has a neighbourhood U in which there exist m complex-valued C^∞ differential 1-forms (remark 2.1.12) $\omega_1, \ldots, \omega_m$ such that the map

$$T_x(V) \to C^m, \qquad x \in U$$

given by

$$X \mapsto (\omega_1(x)(X), \ldots, \omega_m(x)(X))$$

is a C-isomorphism (with respect to the given complex structure on $T_a(V)$).

2.12.1 DEFINITION. An *almost complex structure* on V is a complex structure on each $T_a(V)$ which depends differentiably on a.

The forms $(\omega_1, \ldots, \omega_m)$ whose existence is required are called structure forms. In general, these forms are not *closed*.

For $a \in V$, we denote by J_a the R linear map of $T_a(V)$ into itself given by $X \mapsto \sqrt{-1}\,X$ (multiplication with respect to the given structure of C vector space on $T_a(V)$); for a vector field X, we define the vector field JX by

$$(JX)(a) = J_a X(a).$$

We also denote the almost complex manifold by (V, J) since the map J_a on $T_a(V)$ determines uniquely the structure of C vector space on $T_a(V)$. Of course

$$J_a^2 = -(\text{identity}) \text{ on } T_a(V).$$

We can now apply the considerations of remarks 2.4.10 to $E = T_a(V)$. We have the spaces $\mathscr{E}^* = \mathrm{Hom}_R(E, C)$, $\mathscr{E}^*_{p,q}, \ldots$ In particular,

$$\mathscr{E}^*_{1,0} = \{\omega \in \mathscr{E}^* | \omega(JX) = \mathrm{i}\omega(X) \quad \text{for all} \quad X \in T_a(V)\}.$$

If $(\omega_1, \ldots, \omega_m)$ is a set of structure forms, then $\omega_1(a), \ldots, \omega_m(a)$ lie in $\mathscr{E}^*_{1,0}$ and form a C base of this space. Hence a basis of $\mathscr{E}^*_{p,q}$ is given by the covectors

$$\omega_{j_1}(a) \wedge \ldots \wedge \omega_{j_p}(a) \wedge \bar{\omega}_{k_1}(a) \wedge \ldots \wedge \bar{\omega}_{k_q}(a),$$
$$1 \leqq j_1 < \ldots < j_p \leqq m, \qquad 1 \leqq k_1 < \ldots < k_q \leqq m.$$

We can, as in the case of complex manifolds, speak of differential forms of type (p, q) (definition 2.4.11). On a complex manifold, if ω is a form of type (p, q), then $\mathrm{d}\omega$ is the sum of two forms having types $(p+1, q)$, $(p, q+1)$ respectively. This is no longer generally the case. If $(\omega_1, \ldots, \omega_m)$ is a set of structure forms, then $\mathrm{d}\omega_\nu$ is a 2-form, hence

2.12.2 $$d\omega_v = \eta_v + \eta'_v + \eta''_v,$$

where η_v is of type $(2, 0)$, η'_v of type $(1, 1)$, and η''_v of type $(0, 2)$ we obtain at once:

2.12.3 COROLLARY. If ω is a form of type (p, q) then $d\omega$ is the sum of four forms of types respectively $(p-1, q+2)$, $(p, q+1)$, $(p+1, q)$, $(p+2, q-1)$. Further, the components of type $(p-1, q+2)$, $(p+2, q-1)$ are always zero if and only if the forms η''_v of type $(0, 2)$ in (2.12.2) are zero.

2.12.4 DEFINITION. The almost complex structure on the manifold V is said to be integrable if $d\omega_v$ has no component of type $(0, 2)$ for any set $(\omega_1, \ldots, \omega_m)$ of structure forms.

Let now η be a 2-form, and let X, Y be vector fields. Let

$$S_\eta(X, Y) = \eta(X, Y) + i\eta(JX, Y) + i\eta(X, JY) - \eta(JX, JY).$$

We verify at once the following fact:

2.12.5 COROLLARY. If η is of type $(2, 0)$ or $(1,1)$, we have

$$S_\eta(X, Y) = 0 \qquad \text{for all} \quad X, Y,$$

while, if η is of type $(0, 2)$, we have

$$S_\eta(X, Y) = 2\eta(X, Y).$$

Using the formula of proposition 2.6.6 for a 1-form ω:

$$(d\omega)(X, Y) = X\omega(Y) - Y\omega(X) - \omega([X, Y])$$

and the fact that for any structure form ω_v we have

$$\omega_v(JX) = i\omega_v(X),$$

we obtain easily the following result:

2.12.6 COROLLARY. A complex structure on V is integrable if and only if for any two vector fields X, Y on V, we have

$$[X, Y] + J[JX, Y] + J[X, JY] - [JX, JY] = 0.$$

2.12.7 Remark. We have seen that to any complex analytic structure on V there is associated an almost complex structure (remark 2.4.6), which is integrable (2.6.10). In this case, there is, locally, a set of *closed* structure forms. Let J be the associated mapping of vector fields. A germ of C^∞ functions at $a \in V$ is holomorphic if and only if $(\mathrm{d}f)_x \in \mathscr{E}^*_{1,0}(x)$ for x near a since this means simply that $\bar{\partial} f = 0$ near a. Hence f is holomorphic if and only if

$$(\mathrm{d}f)_x(J_x X) = \mathrm{i}(\mathrm{d}f)_x(X), \qquad X \in T_x(V),$$

i.e., if and only if

2.12.8
$$(J_x X)(f) = \mathrm{i}X(f), \qquad X \in T_x(V).$$

Let now V, V' be complex manifolds, and (V, J), (V', J') the associated almost complex manifolds. Suppose that $f: V \to V'$ is a C^∞ map such that

$$f_{*,a} \circ J_a = J'_{f(a)} \circ f_{*,a} \qquad \text{for all} \quad a \in V.$$

Then f is holomorphic. In fact, we have only to show that if g is a germ of a holomorphic function at $f(a)$, $g \circ f$ is holomorphic at a. But, if $X \in T_a(V)$, we have

$$\begin{aligned}(J_a X)(g \circ f) &= f_{*,a}(J_a X)(g) = (J'_{f(a)} f_{*,a}(X)) \cdot (g)\\ &= \mathrm{i}f_{*,a}(X)(g), \text{ since } g \text{ is holomorphic}\\ &= \mathrm{i}X(g \circ f).\end{aligned}$$

In particular, we deduce:

2.12.9 Proposition. The complex structure underlying an almost complex structure is uniquely determined.

An important theorem of Newlander and Nirenberg [1957] asserts that *any integrable almost complex structure is induced by a complex structure*. Later proofs are due to Kohn [1963] and Hörmander [1964]. Malgrange [1969] has obtained a very simple proof of this result, which is applicable in much more general situations. We do not prove this result. We merely show that if the data are *real analytic*, then the result can be deduced from the Frobenius theorem 2.11.11 for holomorphic forms (see remark 2.11.20).

2.12.10 THEOREM. Let V be a real analytic manifold of dimension $n = 2m$. Suppose that V carries an integrable almost complex structure with the property that, in the neighbourhood of any point, this structure has a set $(\omega_1, \ldots, \omega_m)$ of real analytic structure forms. Then there exists on V, a structure of a complex manifold which underlies both the real analytic and the almost complex structure.

PROOF. Since the theorem is local, we may suppose that V is an open neighbourhood of 0 in R^n, $n = 2m$. We write

$$\omega_\nu = \sum_{\mu=1}^{m} a_{\nu\mu}(x_1, \ldots, x_n)dx_\mu,$$

where the $a_{\nu\mu}$ are complex-valued real-analytic functions in V. By lemma 1.1.5, there is an open set U in $C^n \supset R^n$ and holomorphic functions in U, which we again denote by $a_{\nu\mu}$, extending $a_{\nu\mu}$ from V. Moreover, if we suppose, as we may, that U is a polycylinder, then there are uniquely determined holomorphic function $b_{\nu\mu}$ in U such that

$$b_{\nu\mu}(x) = \overline{a_{\nu\mu}(x)}, \qquad x \in V = U \cap R^n;$$

in fact

$$b_{\nu\mu}(z) = \overline{a_{\nu\mu}(\bar{z})}.$$

We set

$$\eta_\nu = \sum_{\mu=1}^{n} b_{\nu\mu}(z_1, \ldots, z_n)dz_\mu, \qquad \nu = 1, \ldots, m$$

and extend also ω_ν to U by the formula

$$\omega_\nu = \sum_{\mu=1}^{n} a_{\nu\mu}(z_1, \ldots, z_n)dz_\mu, \qquad \nu = 1, \ldots, m.$$

By hypothesis, the $(\omega_1, \ldots, \omega_m)$ form a set of structure forms for the almost complex structure on V. Hence

$$\omega_1(0), \ldots, \omega_m(0), \ \ \bar{\omega}_1(0) = \eta_1(0), \ldots, \bar{\omega}_m(0) = \eta_m(0)$$

form a basis for $\mathscr{E}^* = \mathrm{Hom}_R(T_0(V), C)$. Hence, the ω_ν, η_κ; $1 \leqq \nu, \kappa \leqq m$ are C independent at 0, hence, if U is small enough, they are C independent in U. Hence the

$$dz_\mu, \qquad \mu = 1, \ldots, n$$

are linear combinations in U, with holomorphic coefficients, of the ω_ν, $\eta_\kappa (1 \leqq \nu, \kappa \leqq m)$. Hence the holomorphic 2-forms $d\omega_\nu$ can be written

$$d\omega_\nu = \sum_{r<s} f^{(\nu)}_{r,s} \omega_r \wedge \omega_s + \sum_{r,s} g^{(\nu)}_{r,s} \omega_r \wedge \eta_s + \sum_{r<s} h^{(\nu)}_{r,s} \eta_r \wedge \eta_s,$$

where the coefficients are all holomorphic. Now, by hypothesis, the almost complex structure is integrable. Hence $d\omega_\nu | V$ has no component of type (0, 2), which means that

$$h^{(\nu)}_{r,s} | U \cap R^n = 0.$$

Since $h^{(\nu)}_{r,s}$ is holomorphic, it equals 0 in U, so that $d\omega_\nu$ *is in the ideal generated by the* $\{\omega_\mu\}$ *with coefficients which are holomorphic* 1*-forms.* By the Frobenius theorem (see remark 2.11.20), there are holomorphic functions $F_1, \ldots, F_m$ such that $dF_1, \ldots, dF_m$ generate the same differential system as $\omega_1, \ldots, \omega_m$ in a neighbourhood U' of 0 and $dF_1, \ldots, dF_m$ generate the same subspace of $T^*_z(C^n)$ as $(\omega_1, \ldots, \omega_m)$ for $z \in U'$. Hence $(dF_1, \ldots, dF_m)$ is a set of structure forms of the almost complex structure on $U' \cap R^n$. Moreover, clearly if $\xi_\nu = F_\nu | R^n \cap U'$, then $(d\xi_\nu)(0)$, $(d\bar{\xi}_\nu)(0)$ are R independent and form a basis of $\mathrm{Hom}_R(T_0(R^n), C)$. Hence, the map $\xi: x \mapsto (\xi_1(x), \ldots, \xi_m(x))$ has a differential at 0 which is an isomorphism of R^n onto C^m, so that, by the inverse function theorem (see remark 1.3.11) this is a C^∞ diffeomorphism of a neighbourhood W of 0 in R^n onto an open set W' in C^m. Clearly ξ carries the almost complex structure of W onto that induced by the complex structure of C^m on W'.

The theorem is then an immediate consequence of proposition 2.12.9.

We make one final remark.

2.12.11 REMARK. Any almost complex manifold V carries a natural orientation; in particular, any complex manifold is oriented.

PROOF. We choose a covering $\{U_j\}$ of V such that in each U_j, we have a system $(\omega^{(j)}_1, \ldots, \omega^{(j)}_m)$ of structure forms, $n = 2m$. We set

$$\omega^{(j)} = \left(\frac{i}{2}\right)^m \omega^{(j)}_1 \wedge \overline{\omega^{(j)}_1} \wedge \ldots \wedge \omega^{(j)}_m \wedge \overline{\omega^{(j)}_m}.$$

This is a C^∞ n-form without zeros on U_j. Remark that if

$(\omega_1, \ldots, \omega_m)$, $(\omega'_1, \ldots, \omega'_m)$ are two sets of structure forms on an open set, then there are complex valued functions $a_{\mu\nu}$, $1 \leqq \mu, \nu \leqq n$ such that

$$\omega_\nu = \sum_{\mu=1}^{m} a_{\mu\nu}\, \omega'_\mu .$$

Let $D = \det(a_{\mu\nu})$. Then

$$\omega_1 \wedge \bar{\omega}_1 \wedge \ldots \wedge \omega_m \wedge \bar{\omega}_m = |D|^2 \omega'_1 \wedge \bar{\omega}'_1 \wedge \ldots \wedge \omega'_m \wedge \bar{\omega}'_m,$$

and $|D|^2 > 0$. Hence

$$\omega^{(j)} = f_{jj'}\, \omega^{(j')} \quad \text{on} \quad U_j \cap U_{j'},$$

where $f_{jj'} > 0$. The result follows immediately.

§ 2.13 The lemmata of Poincaré and Grothendieck

We have defined closed and exact forms (definition 2.6.9) and the de Rham groups $H^p(V)$ of a C^∞ manifold V. We prove first a lemma, due to Volterra and usually called Poincaré's lemma, which turns out to be of great importance in the study of these groups.

2.13.1 Poincaré's lemma. Let D be a convex open set in R^n and ω a C^k form ($k \geqq 1$) of degree $p \geqq 1$ on D which is closed, i.e., $d\omega = 0$. Then there is a C^k form ω' of degree $p-1$ on D with $d\omega' = \omega$ and if p is an integer $\geqq 1$, there is an R linear map A: $A^p(D, r) \to A^{p-1}(D, r)$ such that, for any $\omega \in A^p(D, r)$, we have

$$A(d\omega) + dA(\omega) = \omega.$$

A is called a *homotopy operator*. The above result applies to any manifold diffeomorphic with D.

Proof. We may suppose, without loss of generality that the origin 0 of R^n lies in D. Let $I = (0, 1)$ be the open unit interval on R and h: $D \times I$ the map $(t, x) \mapsto tx$.

Let J run through all p-tuples of integers $j_1, \ldots, j_p$ with $1 \leqq j_1 < j_2 < \ldots < j_p \leqq n$ and let $dx_J = dx_{j_1} \wedge \ldots \wedge dx_{j_p}$ if $J = (j_1, \ldots, j_p)$. Any p-form ω on D of class C^k can be written uniquely in the form

$$\omega = \sum_J a_J(x) dx_J, \qquad a_J \in C^k(D).$$

Now $h^*(\omega) = \omega_1 + dt \wedge \omega_0$, where ω_1 is a p-form, ω_0 is a $(p-1)$-form, and neither 'contains dt', i.e., they are of the form

$$\omega_1 = \sum_J b_J(t, x)dx_J, \; J = (j_1, \ldots, j_p), \quad 1 \leqq j_1 < \ldots < j_p \leqq n,$$

$$\omega_0 = \sum_K c_K(t, x)dx_K, \; K = (k_1, \ldots, k_{p-1}), \quad 1 \leqq k_1 < \ldots < k_{p-1} \leqq n.$$

We define a $(p-1)$-form $A(\omega)$ of class C^k on D by

$$A(\omega) = \int_0^1 \omega_0 dt = \sum_K \left(\int_0^1 c_K(t, x)dt \right) dx_K.$$

Clearly A is R linear. Consider now $A(d\omega)$. We have

$$h^*(d\omega) = dh^*(\omega) = d\omega_1 - dt \wedge d\omega_0 = d'\omega_1 + dt \wedge \left(\frac{\partial \omega_1}{\partial t} - d'\omega_0 \right)$$

where, for a p-form χ on $D \times I$, not involving dt, we define $d'\chi$ to be the $(p+1)$-form on $D \times I$, not involving dt, such that

$$d\chi = d'\chi + dt \wedge \frac{\partial \chi}{\partial t}, \qquad \frac{\partial \chi}{\partial t} = \sum \frac{\partial a_J(x, t)}{\partial t} dx_J, \qquad \chi = \sum a_J(x, t)dx_J.$$

Hence

$$A(d\omega) = \int_0^1 \frac{\partial \omega_1}{\partial t} dt - \int_0^1 (d'\omega_0)dt.$$

It is immediately clear that

$$\int_0^1 \frac{\partial \omega_1}{\partial t} dt = \omega$$

and that

$$\int_0^1 d'\omega_0 dt = d \int_0^1 \omega_0 dt = dA(\omega).$$

This gives us

2.13.2 $$\omega = dA(\omega) + A(d\omega).$$

In particular, if ω is closed, $A(d\omega) = 0$, and $\omega = d\omega'$ where $\omega' = A(\omega)$.

There is a corresponding result for the operator $\bar{\partial}$ which was proved first by Grothendieck (unpublished). One has, however, to consider a different class of domains. We begin with the following lemma.

2.13.3 LEMMA. Let K, L, L' be compact sets in C, C^n, R^m respectively. We denote a point in $S = K \times L \times L'$ by (z, w, t). Let g be a C^∞ function in a neighbourhood of S which is holomorphic in w for fixed z, t. Then there is a C^∞ function f in a neighbourhood of S, holomorphic in w for fixed z and t, such that $\partial f/\partial \bar{z} = g$ in a neighbourhood of S.

PROOF. We may suppose that g has compact support in C for fixed w, t; in fact, we have only to multiply, g by a C^∞ function with compact support in a neighbourhood of K, and equal to 1 in a smaller neighbourhood of K.

Let ζ denote any point in C and $\zeta = \xi + i\eta$, ξ, η real. Define

$$f(z, w, t) = -\pi^{-1} \int_C (\zeta - z)^{-1} g(\zeta, w, t) d\xi \wedge d\eta.$$

Since the function $1/\zeta$ is integrable in a neighbourhood of 0 in C, this is equal to

$$-\pi^{-1} \int_C \zeta^{-1} g(z + \zeta, w, t) d\xi \wedge d\eta$$

and so is a C^∞ function, holomorphic in w for fixed z, t. Moreover,

$$\begin{aligned} \frac{\partial f}{\partial \bar{z}}(z, w, t) &= -\pi^{-1} \int_C \zeta^{-1} \frac{\partial g}{\partial \bar{z}}(z + \zeta, w, t) d\xi \wedge d\eta \\ &= \lim_{\varepsilon \to 0} \frac{1}{2\pi i} \int_{|\zeta| \geqq \varepsilon} \frac{\partial g}{\partial \bar{\zeta}}(z + \zeta, w, t) \zeta^{-1} d\bar{\zeta} \wedge d\zeta \\ &= \lim_{\varepsilon \to 0} -\frac{1}{2\pi i} \int_{|\zeta| \geqq \varepsilon} d(g(\zeta + z, w, t) \zeta^{-1} d\zeta). \end{aligned}$$

Now, by Stokes' theorem 2.9.9, this equals

$$\lim_{\varepsilon \to 0} \frac{1}{2\pi i} \int_{|\zeta| = \varepsilon} g(\zeta + z, w, t) \zeta^{-1} d\zeta = g(z, w, t).$$

The result follows.

We shall use the following notation. If ω is a C^∞ form of type (p, q)

in an open set Ω in C^n we can write ω uniquely in the form

$$\omega = \sum_{J,K} a_{JK} dz_J \wedge d\bar{z}_K, \qquad a_{JK} \in C^\infty(\Omega)$$

$$J = (j_1, \ldots, j_p), \quad K = (k_1, \ldots, k_q), \quad 1 \leqq j_1 < \ldots < j_p \leqq n,$$
$$1 \leqq k_1 < \ldots < k_q \leqq n.$$

We call a_{JK} the coefficients of ω. We set

$$\frac{\partial \omega}{\partial \bar{z}_\nu} = \sum_{J,K} \frac{\partial a_{JK}}{\partial \bar{z}_\nu} dz_J \wedge d\bar{z}_K,$$

and define $\partial\omega/\partial z_\nu$ similarly. We then have

$$\bar{\partial}\omega = \sum_{\nu=1}^{n} d\bar{z}_\nu \wedge \frac{\partial \omega}{\partial \bar{z}_\nu}.$$

2.13.4 GROTHENDIECK'S LEMMA. Let $K_1, \ldots, K_m$ be compact sets in C and let $S = K_1 \times \ldots \times K_n \subset C^n$. Let ω be a C^∞ form ω of type (p, q), $q \geqq 1$, in a neighbourhood of S such that $\bar{\partial}\omega = 0$. Then there exists a form ω' of type $(p, q-1)$ in C^n such that $\bar{\partial}\omega' = \omega$ in a neighbourhood of S.

PROOF. If ν is an integer $\geqq 1$, we denote by

$$A_\nu^{p,q} = A_\nu^{p,q}(S)$$

the space of C^∞ forms ω defined in a neighbourhood $U = U(\omega)$ of S and such that ω does not involve $d\bar{z}_\nu, \ldots, d\bar{z}_n$, i.e., ω can be written in the form

$$\omega = \sum a_{JK} \cdot dz_J \wedge d\bar{z}_K, \quad a_{JK} \in C^\infty(U)$$

where $J = (j_1, \ldots, j_p)$, $K = (k_1, \ldots, k_q)$ and $1 \leqq j_1 < \ldots < j_p \leqq n$, $1 \leqq k_1 < \ldots < k_q \leqq \nu - 1$. [If $\nu > n$, this is simply the space of *all* forms of type (p, q).] Clearly, if $\omega \in A_1^{p,q}$ and $q \geqq 1$, we have $\omega \equiv 0$, so that the assertion of lemma 2.13.4 is trivial if $\omega \in A_1^{p,q}$. We suppose, by induction, that the result is proved for all forms in $A_\nu^{p,q}$ where $1 \leqq \nu \leqq n$, and let $\omega \in A_{\nu+1}^{p,q}$. We can write

$$\omega = d\bar{z}_\nu \wedge \omega_1 + \omega_2$$

where

$$\omega_2 \in A_\nu^{p,q}, \qquad \omega_1 \in A_\nu^{p,q-1}.$$

If $\bar{\partial}\omega = 0$, we have

$$-d\bar{z}_\nu \wedge \bar{\partial}\omega_1 + \bar{\partial}\omega_2 = 0.$$

Since ω_1 and ω_2 do not contain $d\bar{z}_\nu, \ldots, d\bar{z}_n$, this implies that

$$\frac{\partial \omega_1}{\partial \bar{z}_j} = 0, \qquad \frac{\partial \omega_2}{\partial \bar{z}_j} = 0$$

for $j \geqq \nu+1$ and that the coefficients of ω_1, ω_2 are holomorphic in $z_{\nu+1}, \ldots, z_n$. By lemma 2.13.3, there exists a form χ' in a neighbourhood of S, of type $(p, q-1)$, all of whose coefficients are holomorphic in $z_{\nu+1}, \ldots, z_n$, such that $\partial\chi'/\partial\bar{z}_\nu = \omega_1$. By multiplying χ' by a suitable C^∞ function with compact support which equals 1 on a neighbourhood of S, we see that there exists a C^∞ form χ on C^n, such that all its coefficients are holomorphic in $z_{\nu+1}, \ldots, z_n$ in a neighbourhood of S and $\partial\chi/\partial\bar{z}_\nu = \omega_1$ in some neighbourhood of S. This implies that

$$\omega - \bar{\partial}\chi \in A_\nu^{p,q}.$$

By our induction assumption on ν there is a $(p, q-1)$ form ψ on C^n with $\omega - \bar{\partial}\chi = \bar{\partial}\psi$ in a neighbourhood of S. This proves the result.

2.13.5 THEOREM. Let $D_1, \ldots, D_n$ be open sets in C and let $D = D_1 \times \ldots \times D_n$. Let ω be a C^∞ form of type (p, q) on D with $q \geqq 1$ and $\bar{\partial}\omega = 0$. Then there is a C^∞ form ω' of type $(p, q-1)$ on D such that $\bar{\partial}\omega' = \omega$.

PROOF. Let $\{K_{\nu,m}\}$, $m = 0, 1, \ldots, \nu = 1, \ldots, n$ be a sequence of compact sets in D_ν such that

$$K_{\nu,m} \subset \mathring{K}_{\nu,m+1}, \qquad \bigcup_{m \geqq 0} K_{\nu,m} = D_\nu,$$

and let $S_m = K_{1,m} \times \ldots \times K_{n,m}$. We consider two cases.

Case I. $q \geqq 2$. By lemma 2.13.4, for any $m \geqq 0$, there is a form ω_m of type $(p, q-1)$ on C^n with $\bar{\partial}\omega_m = \omega$ in a neighbourhood of S_m. Then $\omega_{m+1} - \omega_m$ is a form of type $(p, q-1)$, $q-1 \geqq 1$, with $\bar{\partial}(\omega_{m+1} - \omega_m) = 0$ in a neighbourhood of S_m. Applying lemma 2.13.4

again, there is a form χ_m of type $(p, q-2)$ on C^n with $\bar{\partial}\chi_m = \omega_{m+1}-\omega_m$ on S_m. We define a C^∞ form ω' of type $(p, q-1)$ on D by

$$\omega' = \begin{cases} \omega_0 & \text{on} \quad S_0, \\ \omega_m - \bar{\partial}\chi_{m-1} - \dots - \bar{\partial}\chi_0 & \text{on} \quad S_m, m > 0; \end{cases}$$

note that

$$(\omega_{m+1} - \bar{\partial}\chi_m - \dots - \bar{\partial}\chi_0) - (\omega_m - \bar{\partial}\chi_{m-1} - \dots - \bar{\partial}\chi_0)$$
$$= \omega_{m+1} - \omega_m - \bar{\partial}\chi_m = 0 \quad \text{on} \quad S_m,$$

so that this definition is consistent. Clearly $\bar{\partial}\omega' = \omega$ on D.

Case II. $q = 1$. We suppose now that the sequence $\{K_{\nu, m}\}$ of compact sets has the property that any function holomorphic on a neighbourhood of $K_{\nu, m}$ can be approximated, uniformly on $K_{\nu, m}$, by functions holomorphic on D_ν. It is a classical theorem that such a sequence $\{K_{\nu, m}\}$ exists in any open set in C; this follows also from the results of §3.10. It then follows from theorem 1.7.7 that any function, holomorphic in a neighbourhood of $S_m = K_{1, m} \times \dots \times K_{n, m}$ can be approximated, uniformly on S_m, by functions holomorphic on D.

If $\chi = \sum a_{JK} dz_J \wedge d\bar{z}_K$ is a form on D and $S \subset D$, we write

$$||\chi||_S = \sum_{J, K} \sup_{z \in S} |a_{JK}(z)|.$$

Let ω be a $(p, 1)$ form on D with $\bar{\partial}\omega = 0$. By lemma 2.13.4, there is a form ω_m of type $(p, 0)$ on C^n with $\bar{\partial}\omega_m = \omega$ in a neighbourhood of S_m. Then $\bar{\partial}(\omega_{m+1} - \omega_m) = 0$, so that (see remark 2.6.15) all coefficients of $\omega_{m+1} - \omega_m$ are holomorphic in a neighbourhood of S_m. By our remarks on the possibility of approximation made above, there is a holomorphic $(p, 0)$ form on D, say χ_m, such that

$$||\omega_{m+1} - \omega_m - \chi_m||_{S_m} < 2^{-m}, \qquad m = 0, 1, 2, \dots.$$

We set

$$\omega' = \omega_0 + \sum_{m=0}^{\infty} (\omega_{m+1} - \omega_m - \chi_m).$$

Then ω' is a form of type $(p, 0)$ on D. On S_{m+1}, we have

$$\omega' = \omega_{m+1} - \chi_0 - \dots - \chi_m + \sum_{\mu > m} (\omega_{\mu+1} - \omega_\mu - \chi_\mu).$$

The latter series is a form of type $(p, 0)$ all of whose coefficients are holomorphic in a neighbourhood of S_m, so that, $\chi_0, \ldots, \chi_m$ being holomorphic, we have $\bar{\partial}\omega' = \bar{\partial}\omega_{m+1} = \omega$ on S_m. This being true for any m, the result follows.

The proof of Grothendieck's lemma given above follows the exposition by SERRE [1954] of the original proof of Grothendieck. We remark, moreover, that a similar induction can be used to prove Poincaré's lemma for cubes. This was, in fact, given by V. Volterra; see the Preface to the Third Printing; it was given later by E. CARTAN [1958].

§ 2.14 Applications: Hartogs' continuation theorem and the Oka-Weil theorem

In this section we show how the lemmata of Poincaré and Grothendieck can be applied to complex analysis.

2.14.1 PROPOSITION. Let Ω be a convex open set in C^n and φ a real valued C^2 function on Ω. Then there exists a holomorphic function f on Ω with $\operatorname{Re} f = \varphi$ if and only if $\partial^2\varphi/\partial z_\nu \partial \bar{z}_\mu = 0$ on Ω for $1 \leqq \mu, \nu \leqq n$.

PROOF. If $\varphi = \operatorname{Re} f = \frac{1}{2}(f + \bar{f})$, we have

$$\frac{\partial f}{\partial \bar{z}_\mu} = 0, \qquad \frac{\partial \bar{f}}{\partial z_\nu} = \overline{\frac{\partial f}{\partial \bar{z}_\nu}} = 0,$$

so that $\partial^2\varphi/\partial z_\nu \partial \bar{z}_\mu = 0$. Suppose conversely that these equations are satisfied. This means simply that

$$\bar{\partial}\partial\varphi = 0.$$

Now

$$\mathrm{d}\partial\varphi = (\bar{\partial} + \partial)\partial\varphi = \bar{\partial}\partial\varphi = 0$$

(since $\partial^2 = 0$). Hence by Poincaré's lemma 2.13.1, there exists a complex valued C^1 function g with $\mathrm{d}g = \partial\varphi$. Now $\partial\varphi$ is a form of type $(1, 0)$. Hence $\partial g = \partial\varphi$ and $\bar{\partial} g = 0$ so that g is holomorphic. Further

$$\mathrm{d}(g + \bar{g}) = \partial\varphi + \overline{\partial\varphi} = \mathrm{d}\varphi,$$

(since φ is real) so that $g + \bar{g} - \varphi$ is constant. The result follows.

This proposition implies the following.

2.14.2 COROLLARY. Let φ be a real-valued C^2 function on the complex manifold V. Then φ is locally the real part of a holomorphic function if and only if $\bar{\partial}\partial\varphi = 0$.

2.14.3 LEMMA. Let

$$P = \{(z_1, \ldots, z_n) \in C^n | \ |z_\nu| < R_\nu, \nu = 1, \ldots, n\}$$

be a polycylinder in C^n with $n \geqq 2$. Let U be a neighbourhood of ∂P and f be a function holomorphic in U. Then there exists a holomorphic function F in P and a neighbourhood V of ∂P such that $F|V \cap P = f|V \cap P$.

PROOF. Let $\varepsilon > 0$ and set

$$U_1 = \{(z_1, \ldots, z_n) | \ R_1 - \varepsilon < |z_1| < R_1, |z_\nu| < R_\nu, \nu \geqq 2\},$$

and

$$U_2 = \{(z_1, \ldots, z_n) | \ |z_1| < R_1, R_2 - \varepsilon < |z_2| < R_2, |z_\nu| < R_\nu, \nu \geqq 3\}.$$

Then, if ε is small enough, $U_1 \cup U_2 \subset U$. For any function f holomorphic on U_1, there exist holomorphic functions $a_p(z') = a_p(z_2, \ldots, z_n)$ in

$$\{(z_2, \ldots, z_n) \in C^{n-1} | \ |z_\nu| < R_\nu, \nu = 2, \ldots, n\}$$

such that

$$f(z) = \sum_{p=-\infty}^{\infty} a_p(z')z_1^p, \qquad z = (z_1, z'),$$

and the series converges uniformly on compact subsets of U_1. Choose now any point $z' = (z_2, \ldots, z_n)$ with

$$R_2 - \varepsilon < |z_2| < R_2, \qquad |z_\nu| < R_\nu, \qquad \nu \geqq 3.$$

If f is holomorphic in U, then $f(z_1, z')$ is holomorphic for $|z_1| < R_1$. Hence there are no terms containing negative powers of z_1 in the Laurent expansion of $f(z_1, z')$. This implies that $a_p(z') = 0$ for $p < 0$ if z' satisfies the above condition. By the principle of analytic continuation, $a_p(z') \equiv 0$ if $p < 0$. Hence

$$f(z) = \sum_{p=0}^{\infty} a_p(z')z_1^p \quad \text{on} \quad U_1 \cup U_2.$$

The series converges uniformly on compact subsets of P by Abel's lemma. If

$$F(z) = \sum_{p=0}^{\infty} a_p(z')z_1^p$$

on P, we have $F = f$ in the connected component of $U \cap P$ containing $U_1 \cup U_2$. It is clear that this component is of the form $V \cap P$, where V is a neighbourhood of ∂P.

2.14.4 LEMMA. Let ω be a C^∞ differential form of type (0, 1) in C^n, $n \geqq 2$ having compact support. If $\bar{\partial}\omega = 0$, there exists a C^∞ function φ *with compact support* such that $\bar{\partial}\varphi = \omega$.

PROOF. Choose $R > 0$ such that

$$\text{supp}\,(\omega) \subset P = \{(z_1, \ldots, z_n) \in C^n | \, |z_\nu| < R\}.$$

By Grothendieck's lemma 2.13.4, there is a C^∞ function ψ on C^n with $\bar{\partial}\psi = \omega$. Now, $\bar{\partial}\psi = \omega = 0$ in a neighbourhood U of ∂P, i.e., ψ is holomorphic in a neighbourhood of ∂P. By lemma 2.14.3, there is a holomorphic function F in P extending $\psi | U \cap P$ (if U is suitably chosen). Let

$$\varphi(z) = \begin{cases} \psi(z) - F(z) & \text{if } z \in P, \\ 0 & \text{if } z \notin P. \end{cases}$$

Then φ is a C^∞ function with compact support and $\bar{\partial}\varphi = \omega$.

We now prove the following important theorem due to Hartogs.

2.14.5 THEOREM. Let D be a bounded open set in C^n, $n \geqq 2$ such that $C^n - D$ is connected. Let U be a neighbourhood of ∂D and f be holomorphic in U. Then there exists a neighbourhood V of ∂D and a holomorphic function F on D with $F | V \cap D = f | V \cap D$.

PROOF. Let α be a C^∞ function with compact support in U with $\alpha = 1$ in a neighbourhood of ∂D. Let $f' = \alpha f$ in U, $f' = 0$ in $D - U$. Then $f' \in C^\infty(D)$. Let $\omega = \bar{\partial} f'$ on D, and $\omega = 0$ on $C^n - D$. Since $f' = f$ in a neighbourhood of ∂D and f is holomorphic, $\omega = 0$ in a neighbourhood of ∂D; ω is a C^∞ form with compact support in C^n

(D being bounded). By lemma 2.14.4, there is a C^∞ function φ with compact support in C^n with $\bar\partial\varphi = \omega$. Clearly φ is holomorphic on any open set on which $\omega = 0$, in particular, in a neighbourhood of $C^n - D$. Now $\varphi = 0$ outside a compact subset K of C^n. Since $C^n - D$ is connected, and D is bounded, there is a connected neighbourhood W of $C^n - D$ which meets $C^n - K$. By the principle of analytic continuation, $\varphi \equiv 0$ on W.

Set $F = f' - \varphi$ on D. Since $\varphi = 0$ on W, in particular in a neighbourhood V of ∂D on which $\alpha \equiv 1$, we have $F = f$ on $V \cap D$. Further $\bar\partial F = \bar\partial f' - \bar\partial\varphi = \omega - \bar\partial\varphi = 0$ on D, so that F is holomorphic.

2.14.6 Definition. Let Ω be an open set in C^n. We say that Ω is *$\bar\partial$-acyclic* if for any C^∞ form ω of type (p, q) on Ω with $p \geqq 0$, $q \geqq 1$, $\bar\partial\omega = 0$, there is a C^∞ form ω' of type $(p, q-1)$ such that $\bar\partial\omega' = \omega$.

2.14.7 Theorem (Oka). Let $D = \{z \in C \mid |z| < 1\}$ be the unit disc in C. Let Ω be an open set in C^n such that $\Omega \times D$ is $\bar\partial$-acyclic. Then, if f is holomorphic on Ω, the set

$$\Omega_f = \{x \in \Omega \mid |f(x)| < 1\}$$

is again $\bar\partial$-acyclic. Further, for any C^∞ form ω of type (p, q) on Ω_f, $p \geqq 0$, $q \geqq 0$, $\bar\partial\omega = 0$, there exists a C^∞ form ω' of type (p, q) on $\Omega \times D$ with $\bar\partial\omega' = 0$ such that $u^*(\omega') = \omega$, where $u\colon \Omega_f \to \Omega \times D$ is the map $x \mapsto (x, f(x))$.

Proof. We begin with the proof of the last statement. Let ω be a C^∞ form of type (p, q) on Ω_f, $p, q \geqq 0$, such that $\bar\partial\omega = 0$. Clearly u is a holomorphic, proper, injective map such that u_* is injective at every point. Hence $V = u(\Omega_f)$ is a closed complex submanifold of $\Omega \times D$.

Let $\pi\colon \Omega \times D \to \Omega$ be the projection $(x, z) \mapsto x$ and let $U = \pi^{-1}(\Omega_f) = \Omega_f \times D$. Clearly U is a neighbourhood of V in $\Omega \times D$.

Let U' be a neighbourhood of V in $\Omega \times D$ with $\bar U' \subset U$. Let α be a C^∞ function on $\Omega \times D$ such that

$$\alpha(x, z) = \begin{cases} 0 & \text{if } (x, z) \notin U' \\ 1 & \text{if } (x, z) \text{ is in a neighbourhood of } V. \end{cases}$$

The form $\omega_0 = \pi^*(\omega)$ is a form of type (p, q) on U. Moreover $u^*(\omega_0) = (\pi \circ u)^*(\omega) = \omega$ since $\pi \circ u =$ identity on Ω_f.

Define a form ω_1 on $\Omega \times D$ by

$$\omega_1 = \begin{cases} \alpha\omega_0 & \text{on } U \\ 0 & \text{outside } U. \end{cases}$$

Then ω_1 is a C^∞ form of type (p, q) on $\Omega \times D$ and $u^*(\omega_1) = \omega$. Let

$$\omega_2(x, z) = \begin{cases} 0 & \text{if } (x, z) \in V \\ (z-f(x))^{-1}(\bar{\partial}\omega_1)(x, z) & \text{if } (x, z) \in \Omega \times D - V. \end{cases}$$

Then ω_2 is a C^∞ form of type $(p, q+1)$ on $\Omega \times D$ and $\bar{\partial}\omega_2 = 0$. ($\omega_2$ is C^∞ since $\bar{\partial}\omega_1 = 0$ in a neighbourhood of V.) Since $\Omega \times D$ is $\bar{\partial}$-acyclic, there is a form ω_3 of type (p, q) on $\Omega \times D$ with $\bar{\partial}\omega_3 = \omega_2$.

Let

$$\omega' = \omega_1 - (z-f(x))\omega_3,$$

then ω' is a C^∞ form of type (p, q) on $\Omega \times D$. Moreover

$$\bar{\partial}\omega' = \bar{\partial}\omega_1 - (z-f(x))\omega_2 = 0,$$

and $u^*(\omega') = u^*(\omega_1)$ (since $z-f(x) = 0$ if $(x, z) \in V$) so that $u^*(\omega') = \omega$.

The fact that Ω_f is $\bar{\partial}$-acyclic follows at one from the above result and the fact that $\Omega \times D$ is $\bar{\partial}$-acyclic.

2.14.8 Corollary. With the notation of theorem 2.14.7, if $\Omega \times D^k$ is $\bar{\partial}$-acyclic for every $k \geqq 0$, then so is $\Omega_f \times D^k$ for every $k \geqq 0$.

Proof. This follows from the fact that

$$\Omega_f \times D^k = \Omega'_g,$$

where $\Omega' = \Omega \times D^k$, and $g(x, z) = f(x)$.

2.14.9 Theorem (Oka). Let $f_1, \ldots, f_k$ be holomorphic functions on C^n and let

$$U = \{x \in C^n |\; |f_\nu(x)| < 1, \nu = 1, \ldots, k\}.$$

Let $u: U \to C^n \times D^k$ be the map

$$x \mapsto (x, f_1(x), \ldots, f_k(x)).$$

Then, for any holomorphic function g on U, there is a holomorphic function G on $C^n \times D^k$ such that $G \circ u = g$.

PROOF. Let

$$\Omega_0 = C^n, \qquad \Omega_p = \{x \in \Omega_{p-1} |\ |f_p(x)| < 1\}, \qquad 1 \leqq p \leqq k.$$

Let

$$u_p \colon \Omega_{k-p} \times D^p \to \Omega_{k-p-1} \times D^{p+1}$$

be the map

$$(x, z) \mapsto (x, z, f_{k-p}(x)).$$

By theorem 2.13.5 and corollary 2.14.8, for $0 \leqq r \leqq k$, $s \geqq 0$, $\Omega_r \times D^s$ is $\bar{\partial}$-acyclic. Moreover, by theorem 2.14.7, for any $\bar{\partial}$-closed (p, q) form ω on $\Omega_{k-p} \times D^p$, there is a $\bar{\partial}$-closed (p, q) form ω' on $\Omega_{k-p-1} \times D^{p+1}$ such that $u_p^*(\omega') = \omega$.

Now, we have $\Omega_k = U$ and

$$u \colon U \to C^n \times D^k$$

is the map

$$u = u_{k-1} \circ \ldots \circ u_1 \circ u_0 .$$

It follows that for any form ω of type (p, q) on U with $\bar{\partial}\omega = 0$, there is a form ω' of type (p, q) on $\Omega_0 \times D^k = C^n \times D^k$ with $\bar{\partial}\omega' = 0$ and $u^*(\omega') = \omega$; theorem 2.14.9 is the case $p = q = 0$ of this statement.

2.14.10 OKA-WEIL APPROXIMATION THEOREM. If $f_1, \ldots, f_k$ are holomorphic functions on C^n and if

$$U = \{x \in C^n |\ |f_\nu(x)| < 1, \nu = 1, \ldots, k\},$$

then U is a Runge domain (see definition 1.7.1), i.e. any holomorphic function on U can be approximated by polynomials in $z_1, \ldots, z_n$, uniformly on compact subsets of U.

PROOF. Let u: $U \to C^n \times D^k = \Omega$ be the map $x \mapsto (x, f_1(x), \ldots, f_k(x))$. If g is holomorphic on U, there exists, by theorem 2.14.9, a holomorphic function G on Ω with $G \circ u = g$. Now G can be expanded in a Taylor series

$$G(x, z) = \sum a_{\alpha\beta} x_1^{\alpha_1} \ldots x_n^{\alpha_n} z_1^{\beta_1} \ldots z_k^{\beta_k},$$

which converges uniformly on compact subsets of Ω; in particular, G is the limit of polynomials P_N which converge uniformly on compact sets in Ω. Hence the entire functions

$$g_N(x) = P_N(x_1, \ldots, x_n, f_1(x), \ldots, f_k(x))$$

converge uniformly to $G \circ u = g$ on compact subsets of U. Since any entire function is the limit of polynomials, the theorem follows.

2.14.11 PROPOSITION. A convex open set in C^n is a Runge domain.

PROOF. It is sufficient to prove that a bounded convex open set Ω in C^n is a Runge domain. Let K be a compact subset of Ω. If $x_1, \ldots, x_n$ are the coordinate functions in C^n, for any boundary point $a \in \partial\Omega$, there is a linear function

$$l_a(x) = \sum_{\nu=1}^{n} c_\nu x_\nu + c_0,$$

such that $l_a(a) = 0$ and $\operatorname{Re} l_a(x) < 0$ for any $x \in \Omega$. It follows that for any $a \in \partial\Omega$, there is a linear function L on C^n such that

$$\operatorname{Re} L(x) < 0, \qquad x \in K, \qquad \operatorname{Re} L(a) > 0,$$

(replace l_a by $L = l_a + \delta$ where $\delta > 0$ is sufficiently small). Then $\operatorname{Re} L(x) > 0$ for all x near a. Since $\partial\Omega$ is compact, there exist finitely many linear functions $L_1, \ldots, L_r$ such that

$$\max_\nu \operatorname{Re} L_\nu(x) \begin{cases} > 0 & \text{if } x \in \partial\Omega, \\ < 0 & \text{if } x \in K. \end{cases}$$

Hence the set

$$U = \{x \in \Omega |\ \operatorname{Re} L_\nu(x) < 0, \nu = 1, \ldots, r\}$$

contains K and is relatively compact in U. Now the set $V = \{x \in C^n |\ \operatorname{Re} L_\nu(x) < 0\}$ is convex in C^n, hence connected, and $V \cap \Omega = U$ is relatively compact in Ω. Hence $V = U$. Thus

$$U = \{x \in C^n |\ |f_\nu(x)| < 1, f_\nu = \exp(L_\nu), \nu = 1, \ldots, r\}.$$

By theorem 2.14.10, U is a Runge domain, so that holomorphic functions on U can be approximated, uniformly on K, by polynomials. Since K is any compact set in Ω and $U \subset \Omega$, the proposition follows.

The proof of Hartogs' theorem given here is suggested by the proof of the approximation theorem of Malgrange-Lax (see § 3.10; also MALGRANGE [1956]). That of the Oka-Weil theorem is essentially a translation of OKA'S own proof [1936] into the language of differential forms. Both are given also in HÖRMANDER [1966].

§ 2.15 Immersions and imbeddings: Whitney's theorems

We shall deal only with C^k manifolds $V, V', \ldots$ which are countable at infinity.

2.15.1 DEFINITION. Let V, V' be C^k manifolds and $f: V \to V'$ a C^k map. f is called *regular* at $a \in V$ if the differential map $f_{*,a}: T_a(V) \to T_{f(a)}(V')$ is injective. f is called regular on a set $S \subset V$ if it is regular at every $a \in S$; it is simply called regular, if it is regular on V.
A C^k map $f: V \to V'$ is called an imbedding if f is an immersion and injective. If f is, in addition, a homeomorphism of V onto $f(V)$ with the topology induced from V', f is called a locally proper imbedding (see proposition 2.5.8). An imbedding (immersion) $f: V \to V'$ is called a closed imbedding (immersion) if f is proper (see definition 2.5.7). Similar definitions apply to real and complex analytic manifolds and maps.

We now consider the case when $V' = R^q$. Let $C^k(V, q)$ denote the set of all C^k maps of V into R^q. We topologize this space as follows:
Let $\mathfrak{U}$ be a family of coordinate systems

$$\mathfrak{U} = \{U_i, \varphi_i\}_{i \in \mathscr{I}}$$

such that the $\{U_i\}$ form a locally finite covering of V. We suppose that $U_i \Subset V$. Let K_i be a compact subset of U_i such that $\bigcup K_i = V$. We consider a family $\varepsilon = \{\varepsilon_i\}_{i \in \mathscr{I}}$ of positive real numbers and a family $\mathfrak{m} = \{m_i\}_{i \in \mathscr{I}}$ of positive integers $m_i \leqq k$, both indexed by the same set $\mathscr{I}$ as $\mathfrak{U}$. For any C^k function ψ on U_i and an n-tuple of integers $\alpha = (\alpha_1, \ldots, \alpha_n)$, $\alpha_\nu \geqq 0$, where $|\alpha| = \alpha_1 + \ldots + \alpha_n \leqq k$, we set

$$D^\alpha\psi(x) = \left(\frac{\partial}{\partial x_1}\right)^{\alpha_1} \cdots \left(\frac{\partial}{\partial x_n}\right)^{\alpha_n} \psi(x),$$

where the vector fields $\partial/\partial x_\nu$ corresponding to the coordinate system $\{U_i, \varphi_i\}$ are defined as in § 2.4. For $f_0 \in C^k(V, q)$, let

$$\mathscr{B} = \mathscr{B}(\mathfrak{U}, \mathfrak{m}, \varepsilon, f_0)$$

be the set

$$\mathscr{B} = \{f \in C^k(V, q) |\ |D^\alpha(f - f_0)(x)| < \varepsilon_i \quad \text{for} \quad x \in K_i, |\alpha| \leqq m_i, i \in \mathscr{I}\}.$$

We define the topology on $C^k(V, q)$ by the requirement that the sets $\mathscr{B}$ with $\mathfrak{U}$ fixed, when $\mathfrak{m}$, ε run over all the families having the properties listed above, form a fundamental system of neighbourhoods of f_0. If

$$\mathfrak{V} = \{V_j, \varphi'_j\}_{j \in J}$$

is another family of coordinate systems as above, and $\varepsilon' = \{\varepsilon'_j\}$, $\mathfrak{m}' = \{m'_j\}$ correspond to $\mathfrak{V}$, one sees at once that for any set

$$\mathscr{B}(\mathfrak{V}, \mathfrak{m}', \varepsilon', f_0) = \mathscr{B}'$$

there exists a set $\mathscr{B}(\mathfrak{U}, \mathfrak{m}, \varepsilon, f_0)$ contained in $\mathscr{B}'$ and conversely. Hence the topology on $C^k(V, q)$ is independent of the family $\mathfrak{U}$ (and the system of compact sets $K_i \subset U_i$) chosen.

We denote the space $C^k(V, q)$ with this topology by $\mathfrak{C}^k(V, q)$ or just $\mathfrak{C}^k$ or $\mathfrak{C}^k(V)$ when no confusion is likely.

Note that this topology coincides with the 'topology of uniform convergence on compact sets of derivatives of order $\leqq k$' (see § 2.16) only in the case when V is compact. $\mathfrak{C}^k(V, q)$ does not have, in general, a countable base and is not metrisable.

The following proposition follows easily from the definitions:

2.15.2 Proposition. The set of C^k immersions $f\colon V \to R^q$ is open in $\mathfrak{C}^k(V, q)$ for $k \geqq 1$.

We fix a covering $\mathfrak{U} = \{U_i, \varphi_i\}$ with the properties listed earlier. If L is any compact set in V, and $f \in C^k(V, q)$, we set, for $0 \leqq r \leqq k$,

$$||f||_{r,L} = \sum_i \sum_{|\alpha| \leqq r} \sup_{x \in L \cap K_i} |D^\alpha f(x)|.$$

2.15.3 LEMMA. Let K be a compact set in V and L a compact neighbourhood of K. Then for any C^k imbedding $f: V \to R^q$, there exists a $\delta > 0$ such that for all $g \in C^k(V, q)$ with

$$||f-g||_{1,L} < \delta,$$

the map $g|K$ is injective.

PROOF. Let (U, φ) be a coordinate system on V such that U is relatively compact and let $C \subset U$ be compact. If $a, b \in U$, we write

$$d(a, b) = ||\varphi(a)-\varphi(b)||.$$

Then, there is an $\varepsilon > 0$ such that

$$|f(a)-f(b)| \geqq \varepsilon d(a, b), \qquad a, b \in C,$$

since f is an imbedding. If $g \in C^k(V, q)$ and $||f-g||_{1,\bar{U}}$, is sufficiently small, then, for $h = f-g$, we have

$$|h(a)-h(b)| \leqq \tfrac{1}{2}\varepsilon d(a, b) \qquad \text{for} \quad a, b \in C.$$

If follows that $g|C$ is injective; in fact $|g(a)-g(b)| \geqq \frac{1}{2}\varepsilon d(a, b)$ for $a, b \in C$.

Since K is compact, we can choose coordinate systems $\{V_1, \varphi_1\}, \ldots,$ $\{V_N, \varphi_N\}$ with $K \subset \bigcup V_\nu \subset L$. If $||f-g||_{1,L}$ is sufficiently small, then $g|V_\nu$ is injective. Hence there exists a neighbourhood W of the diagonal Δ of $K \times K$ such that if $(a, b) \in W-\Delta$, $g(a) \neq g(b)$ for any g with $||f-g||_{1,L}$ small.

Now, there is a $\theta > 0$ such that $|f(a)-f(b)| \geqq \theta$ if $(a, b) \in K \times K-W$. If $||f-g||_{0,K} < \frac{1}{2}\theta$, then for $(a, b) \in K \times K-W$: $|g(a)-g(b)| > 0$. The lemma follows at once.

2.15.4 PROPOSITION. The set of closed imbeddings of V in R^q is open in $\mathfrak{C}^k(V, q)$ if $k \geqq 1$.

PROOF. Let $\{K_m\}$ be a sequence of compact sets in V such that

$$K_0 = \emptyset, \qquad K_m \subset \mathring{K}_{m+1} \quad \text{and} \quad \bigcup K_m = V.$$

Let L_m be the closure of $K_{m+1} - K_m$. Then

$$L_m \cap L_{m'} = \emptyset' \qquad \text{if} \quad m' \geqq m+2$$

and the $\{L_m\}$ form a locally finite family. If $f\colon V \to R^q$ is a closed imbedding, the same is then true of the sets $f(L_m) \subset R^q$. Hence there exist open sets U_m in R^q, such that $f(L_m) \subset U_m$ and $U_m \cap U_{m'} = \emptyset$ if $m' \geqq m+2$. We can choose $\delta_m > 0$ such that if $g \in C^k(V, q)$ and

$$\|f-g\|_{1, L_m} < \delta_m$$

for all $m > 0$, then $g(L_m) \subset U_m$ and $g|L_m \cup L_{m+1}$ is injective (by lemma 2.15.3). We claim that all such g are injective. In fact, if $a, b \in V$, $a \neq b$, let $a \in L_m$, $b \in L_{m'}$ (with, say, $m' \geqq m$). If $m' \geqq m+2$, then $g(a) \in U_m$, $g(b) \in U_{m'}$, and $U_m \cap U_{m'} = \emptyset$ so that $g(a) \neq g(b)$. If $m' = m+1$ or m, then $g(a) \neq g(b)$ since $g|L_m \cup L_{m+1}$ is injective. We deduce at once from proposition 2.15.2 that there is a neighbourhood $\mathscr{B}$ of f such that any $g \in \mathscr{B}$ is an imbedding. Trivially, if $f\colon V \to R^q$ is proper and $|f(x)-g(x)| \leqq 1$ for all $x \in V$, then g is also proper. Proposition 2.15.4 follows.

2.15.5 REMARK. It is easily seen as in the above proof that any g in a suitable neighbourhood of a locally proper imbedding is again an imbedding. Without the assumption that the imbedding be locally proper, the statement is false. Further, even on compact sets, an approximation to a (non-regular) injective map need not be injective.

2.15.6 LEMMA. Let Ω be a bounded open set in R^n and $f\colon \Omega \to R^q$ a C^k map, $k \geqq 1$, $q \geqq 2n$. Let r be any integer $\leqq k$. Then for any $\varepsilon > 0$, there exists a C^k map $g\colon \Omega \to R^q$ such that $\|g-f\|_r^\Omega < \varepsilon$ and such that the vectors $\partial g/\partial x_1, \ldots, \partial g/\partial x_n$ are linearly independent at every point of Ω.

PROOF. We may suppose that $k \geqq 2$, by theorem 1.6.5. Let $f_0 = f$. It is sufficient to show that if f' is a C^k map such that $\partial f'/\partial x_1, \ldots, \partial f'/\partial x_s$ are linearly independent at any point of Ω, then there is

$g \in C^k(\Omega)$ with $\|g-f'\|_r^\Omega < \varepsilon$ such that $\partial g/\partial x_1, \ldots, \partial g/\partial x_{s+1}$ are independent at any point of Ω; here $0 \leqq s < n$. Let

$$v_j(x) = \frac{\partial f'}{\partial x_j}, \qquad 1 \leqq j \leqq n;$$

v_j is a C^{k-1} map from Ω to R^q. Consider the map

$$\varphi: R^s \times \Omega \to R^q$$

defined by

$$\varphi(\lambda_1, \ldots, \lambda_s, x) = \sum_{j=1}^{s} \lambda_j \frac{\partial f'}{\partial x_j} - v_{s+1}(x).$$

Now, $\dim (R^s \times \Omega) = s+n < 2n \leqq q$ and $\varphi \in C^1$ (since $k \geqq 2$). Hence, by lemma 1.4.3, for any $\delta > 0$, there is an $a \in R^q$, $\|a\| < \delta$, such that $a \notin \varphi(R^s \times \Omega)$. If we set

$$g(x) = f'(x) + a \cdot x_{s+1},$$

then, if $\partial f'/\partial x_1, \ldots, \partial f'/\partial x_s$ are R independent and $a \notin \varphi(R^s \times \Omega)$, then $\partial g/\partial x_1, \ldots, \partial g/\partial x_{s+1}$ are R independent at any $x \in \Omega$. If δ is small enough, we obtain our result.
Note that the g of lemma 2.15.6 is an immersion of Ω.

2.15.7 Whitney's immersion theorem. *If V is a C^k manifold of dimension n and $q \geqq 2n$, then the set of C^k immersions of V into R^q is an open dense set in $\mathfrak{C}^k(V, q)$.*

Proof. Let $\mathfrak{U} = \{U_\nu, \varphi_\nu\}$, $\nu = 1, 2, \ldots$ be a sequence of coordinate systems such that $\{U_\nu\}$ is locally finite and $\varphi_\nu(U_\nu) = \Omega_\nu$ is a bounded open set in R^n, U_ν is relatively compact in V and $V = \bigcup U_\nu$. Let $K_\nu \subset U_\nu$ be compact, $\bigcup K_\nu = V$ and let $\varepsilon_\nu > 0$ and integers $n_\nu \geqq 0$, $n_\nu \leqq k$ be given. Let $f \in C^k(V, q)$, and set $f_0 = f$. Assume that $f_0, \ldots, f_m$ have been constructed having the following properties:

(a, m) $|D^\alpha f_m(x) - D^\alpha f(x)| < \varepsilon_\nu$ for $x \in K_\nu$, $|\alpha| \leqq n_\nu$;
(b, m) f_m is regular on $\bigcup_{\nu=0}^m K_\nu$;
(c, m) $\operatorname{supp}(f_{p+1} - f_p) \subset U_{p+1}$, $p = 0, 1, \ldots, m-1$.

Let α_m be a C^k function on V with support in U_{m+1} which equals 1 in a neighbourhood of K_{m+1}. By lemma 2.15.6, there is, for any $\delta > 0$ and any integer $0 \leqq N \leqq k$ a C^k map $h_m: U_{m+1} \to R^p$ which is regular on U_{m+1} and

$$|D^\alpha(f_m - h_m)| < \delta \quad \text{on} \quad U_{m+1}, \qquad |\alpha| \leqq N.$$

Consider the map

$$f_{m+1} = \begin{cases} f_m + \alpha_m(h_m - f_m) & \text{on} \quad U_{m+1}, \\ f_m & \text{on} \quad V - U_{m+1}. \end{cases}$$

Then $f_{m+1} \in C^k(V, q)$ and, as $\delta \to 0$ and $N \to k, f_{m+1} \to f_m$ in $\mathfrak{C}^k(V, q)$. Hence, if δ is small enough and $N \geqq 1, f_{m+1}$ is regular on $\bigcup_{\nu \leqq m} K_\nu$. Further, $f_{m+1} = h_m$ on K_{m+1}, and so is regular on K_{m+1}. Clearly, if δ is small enough and N is large enough, the maps $(f_0, \ldots, f_{m+1})$ satisfy $(a, m+1)$, $(b, m+1)$, $(c, m+1)$. By induction, we have maps $f_m: V \to R^q$, $m \geqq 0$, satisfying (a, m), (b, m), (c, m) for all $m \geqq 0$. If we define

$$g = \lim_{m \to \infty} f_m,$$

we have a regular map $g: V \to R^q$ with $|D^\alpha f(x) - D^\alpha g(x)| \leqq \varepsilon_\nu$, $|\alpha| \leqq n_\nu$ for $x \in K_\nu$. This proves the theorem.

2.15.8 Whitney's imbedding theorem. Let V be a C^k manifold of dimension n and $q \geqq 2n+1$. Then the set of imbeddings of V into R^q is dense in $\mathfrak{C}^k(V, q)$.

Proof. Let f be any immersion of V into R^q. Let $\mathfrak{U} = \{U_\nu, \varphi_\nu\}$ be a sequence of coordinate neighbourhoods of V such that $\{U_\nu\}$ is a locally finite covering of V, $U_\nu \Subset V$, and $f|U_\nu$ is injective (such a sequence exists). Let $\varepsilon_\nu > 0$ and $n_\nu > 0$ be integers $\leqq k$. Let K_ν be compact in U_ν and $\bigcup K_\nu = V$. Let $f_0 = f$. We shall construct, by induction, mappings $f_0, \ldots, f_m$ of V into R^q such that the following properties hold:

(i, m) $f_m|U_\nu$ is injective for all ν, $f_m|\bigcup_{\nu \leqq m} K_\nu$ is injective;

(ii, m) $\operatorname{supp}(f_{p+1} - f_p) \subset U_{p+1}$ for $p = 0, 1, \ldots, m-1$;

(iii, m) $|D^\alpha f_m(x) - D^\alpha f(x)| < \varepsilon_\nu$ for $x \in K_\nu$ and $|\alpha| \leqq n_\nu$, $\nu \geqq 1$.

Suppose $f_0, \ldots, f_m$ given. Let α_m be a C^k function on V with compact support in U_{m+1}, $\alpha_m = 1$ in a neighbourhood of K_{m+1}. Consider the set

$$\Omega = \{(x, y) \in V \times V | \alpha_m(x) \neq \alpha_m(y)\};$$

Ω is a C^k manifold of dimension $2n$. Let $\varphi\colon \Omega \to R^q$ be the map

$$\varphi(x, y) = -\frac{f_m(x) - f_m(y)}{\alpha_m(x) - \alpha_m(y)}.$$

Since $q \geqq 2n+1$, $\varphi(\Omega)$ has measure zero; hence, for any $\delta > 0$, there is a

$$v \in R^q, \qquad \|v\| < \delta, \qquad v \notin \varphi(\Omega).$$

Set

$$f_{m+1} = f_m + \alpha_m v.$$

We claim that f_{m+1} has properties (i, $m+1$) and (ii, $m+1$); the latter is obvious. Now if

$$f_{m+1}(x) = f_{m+1}(y)$$

we have

$$f_m(x) - f_m(y) = -v(\alpha_m(x) - \alpha_m(y)),$$

so that, since $v \notin \varphi(\Omega)$, we must have $\alpha_m(x) - \alpha_m(y) = 0$, and $f_m(x) - f_m(y) = 0$. The latter fact implies that $f_{m+1}|U_\nu$ is injective for all ν and that $f_{m+1}|\bigcup_{\nu \leqq m} K_\nu$ is injective. Let $x \in K_{m+1}$ and $y \in \bigcup_{\nu \leqq m+1} K_\nu$ and suppose that $f_{m+1}(x) = f_{m+1}(y)$. Then since $\alpha_m(x) = \alpha_m(y)$, it follows that $y \in U_{m+1}$ (since $\alpha_m(x) = 1$ since $x \in K_{m+1}$ and $\alpha_m = 0$ on $V - U_{m+1}$). But since $f_m|U_{m+1}$ is injective, this implies that $x = y$.

Moreover, when $\delta \to 0$, $f_{m+1} \to f_m$ in $\mathfrak{C}^k(V, q)$. The existence of maps f_m satisfying (i, m), (ii, m), (iii, m) for all $m \geqq 0$ follows easily.

If we set

$$g = \lim_{m \to \infty} f_m,$$

then g is injective on V, and we have $|D^\alpha(f-g)| \leqq \varepsilon_\nu$ for $|\alpha| \leqq n_\nu$ on K_ν. This shows that arbitrarily close to any immersion, there exists a C^k injection of V into R^q. The theorem follows from theorem 2.15.7 and proposition 2.15.2.

From these theorems, and proposition 2.15.4 we deduce immediately:

2.15.9 THEOREM. Let V be a C^k manifold of dimension n, $k \geqq 1$. If $q \geqq 2n$, the set of closed immersions of V in R^q is an open dense subset of the open set $\mathfrak{P}$ of all proper C^k maps of V into R^q. If $q \geqq 2n+1$, the set of closed imbeddings of V in R^q is open and dense in $\mathfrak{P}$.

2.15.10 REMARK. The set $\mathfrak{P}$ of proper C^k maps of V into R^q ($q \geqq 1$), is non-empty.

PROOF. Since there is a proper C^k map of R into R^q ($q \geqq 1$), we may suppose that $q = 1$. Let $\{U_\nu\}$ be a sequence of relatively compact subsets of V, $\bigcup U_\nu = V$, and suppose that the family $\{U_\nu\}$ is locally finite. Let K_ν be a compact subset of U_ν such that $\bigcup K_\nu = V$. Let η_ν be a C^k function on V with compact support in U_ν so that $0 \leqq \eta_\nu(x) \leqq 1$ for all x, $\eta_\nu(x) = 1$ for $x \in K_\nu$. Then the map

$$\varphi : V \to R$$

given by

$$\varphi(x) = \sum_{\nu \geqq 1} \nu \eta_\nu(x)$$

is proper and C^k.

2.15.11 COROLLARY. If V is a C^k manifold of dimension n, there is a closed immersion of V in R^{2n} and a closed imbedding of V in R^{2n+1}.

2.15.12 REMARKS. We add here a few remarks on the imbedding and immersion theorems of Whitney. In the first place, the restriction of the second manifold to euclidean space is not essential. One can prove the following theorem (see e.g. CARTAN [1962]).

Let V and V' be C^k manifolds (with countable base) and let V' be connected. Suppose that $\dim V' \geqq 2 \dim V+1$. Then there is a closed C^k imbedding of V in V' unless V is noncompact and V' is compact. [In the latter case, there is of course no proper map of V into V'.]

Now, when V is a real analytic manifold, and if V admits a real analytic closed imbedding in R^q for some q, then, it follows easily from proposition 2.5.14 and theorem 1.6.5, that *real analytic functions on V are dense in* $\mathfrak{C}^k(V, 1)$. Hence, from the results of Whitney proved above and proposition 2.15.4, it follows that such a manifold has a real analytic closed immersion in R^{2n} and a closed analytic imbedding in R^{2n+1}. These results have been completed by GRAUERT [1958] who has shown that any real analytic manifold with a countable base (of dimension n) can be real analytically imbedded as a closed submanifold in R^q for some q.

Returning to C^k manifolds, WHITNEY [1944b] has sharpened the immersion theorem; he has shown that any C^k manifold of dimension $n \geqq 2$ has a closed C^k immersion in R^{2n-1}. [This is obviously false for $n = 1$; the circle cannot be immersed in the line.] He has further proved [1944a] that any C^k manifold of dimension n can be imbedded in R^{2n}; in particular, compact C^k manifolds of dimension n, have closed imbeddings in R^{2n}. These results have been completed by HIRSCH [1961] who has shown that *a non-compact manifold of dimension n has an imbedding in* R^{2n-1}, *hence a closed imbedding in* R^{2n}.

Putting these remarks together, one can in particular assert the following:

Theorem. Let V be a C^k (real analytic) manifold of dimension n. Then there is a closed C^k (real analytic) immersion of V in R^{2n-1} (if $n \geqq 2$) and a closed C^k (real analytic) imbedding of V in R^{2n}.

Very interesting results which prove sharper imbedding theorems for more restricted classes of manifolds are known. These 'imbedding and non-imbedding' theorems have given rise to an extensive literature. We shall content ourselves with stating two of these theorems. Further references may be found in ATIYAH [1962] and HAEFLIGER [1961].

Theorem of Wall (see WALL [1965]). Any compact manifold of dimension 3 can be imbedded in R^5.

A manifold is called k-connected if, for any m with $0 \leqq m \leqq k$ and any continuous map f of the sphere

$$S^m = \{(x_1, \ldots, x_{m+1}) \in R^{m+1} | x_1^2 + \ldots + x_{m+1}^2 = 1\}$$

into V, there is a continuous map F of the disc

$$D^{m+1} = \{(x_1, \ldots, x_{m+1}) \in R^{m+1} | x_1^2 + \ldots + x_{m+1}^2 \leqq 1\}$$

into V such that $F|S^m = f$.

Theorem of Haefliger [1961]: If V is a compact k-connected manifold, then V can be imbedded in R^{2n-k} if $n \geqq 2k+3$.

The problem of imbedding complex manifolds holomorphically in some C^q is of a quite different nature. The manifolds that can be imbedded as closed submanifolds of C^q are the so called *Stein manifolds*. For these, one has the analogue of corollary 2.15.11, see BISHOP [1961] and NARASIMHAN [1960].

Note. The proof of Theorems 2.15.7,8 given here are due essentially to WHITNEY [1957].

§ 2.16 Thom's transversality theorem

In this section we prove a special case of a theorem of THOM [1956]. The proof we give is essentially that of ABRAHAM [1963]. For a closer study of the theorem and some applications, see CARTAN [1962].

Let V be a C^k manifold of dimension n, $1 \leqq k \leqq \infty$, with a countable base. On the space $C^k(V)$ of C^k functions (with real values) on V we define a topology as follows. Let K be any compact subset of V, $X_1, \ldots, X_m$ $(0 \leqq m \leqq k,\ m \in Z)$ vector fields on V and $\varepsilon > 0$. The sets

$$\mathscr{B} = \{f \in C^k(V) |\ |X_1 \ldots X_m(f)(x)| < \varepsilon \quad \text{for all } x \in K\}$$

form a fundamental system of neighbourhoods of $0 \in C^k(V)$. (This topology of 'convergence of derivatives of order $\leqq k$ uniformly on compact sets' is weaker than the topology $\mathfrak{C}^k$ introduced in § 2.15.) Let V' be a C^k manifold of dimension m with a countable base and $C^k(V, V')$ the set of C^k maps of V into V'. We topologize $C^k(V, V')$ as follows.

A filter $\{f_\alpha\}$ of C^k maps f_α: $V \to V'$ converges to a C^k map f: $V \to V'$ if and only if, for every C^k function φ on V', the filter $\{\varphi \circ f_\alpha\}$ of C^k functions on V converges in $C^k(V)$.

It is easily verified that $C^k(V, V')$ is a complete metric space with a countable base.

Let now V, V' be two C^k manifolds countable at infinity, and let (W, i) be a closed submanifold of V'. We identify W with $i(W)$ and the tangent space $T_a(W)$ at $a \in W$ with the subspace $i_{*,a}(T_a(W))$ of $T_{i(a)}(V')$.

2.16.1 DEFINITION. A C^k map $f: V \to V'$ is said to be *transversal* to W at a point $a \in V$ if either $f(a) \notin W$ or

$$f_{*,a}(T_a(V)) + T_{f(a)}(W) = T_{f(a)}(V'),$$

i.e., if $f_{*,a}(T_a(V))$ and $T_{f(a)}(W)$ span $T_{f(a)}(V')$. f is said to be transversal to W if it is transversal to W at every point of V.

In what follows, $\dim V = n$, $\dim V' = m$ and $\dim W = m-q$, $q \geqq 1$, so that W has codimension q.

2.16.2 PROPOSITION. If $f: V \to V'$ is a C^k map and $f(a) \in W$, $a \in V$, then f is transversal to W if and only if there is a subspace E of $T_a(V)$ of dimension q such that $f_{*,a}|E$ is injective and

$$f_{*,a}(E) \cap T_{f(a)}(W) = \{0\}.$$

PROOF. If the above condition is satisfied, since $\dim f_{*,a}(E) + \dim T_{f(a)}(W) = \dim T_{f(a)}(V')$, we have $f_{*,a}(E) + T_{f(a)}(W) = T_{f(a)}(V')$, in particular, f is transversal to W at a. Conversely, if f is transversal to W at a, let E' be a subspace of $f_{*,a}(T_a(V))$ of dimension q such that $E' \cap T_{f(a)}(W) = \{0\}$. There is a subspace E of $T_a(V)$ of dimension q such that $f_{*,a}(E) = E'$ and the result follows.

2.16.3 PROPOSITION. If $f: V \to V'$ is a C^k map transversal to W, then $f^{-1}(W)$ is either empty or a submanifold of V of codimension q, i.e., of dimension $n-q$ (the inclusion of $f^{-1}(W)$ in V makes it a submanifold).

PROOF. Let $a \in f^{-1}(W)$ and $b = f(a) \in W$. Let E be a q-dimensional subspace of $T_a(V)$ such that $f_{*,a}|E$ is injective and $E' \cap T_b(W) = \{0\}$, where $E' = f_{*,a}(E)$. Let $g_1, \ldots, g_q$ be C^k functions in a neighbourhood U' of b on V' such that $dg_1, \ldots, dg_q$ are R independent and

$$U' \cap W = \{y \in U' | g_1(y) = \ldots = g_q(y) = 0\}.$$

Since $(dg_j)_b | T_b(W) = 0$, it follows that $(dg_1)_b|E', \ldots, (dg_q)_b|E'$ are R independent. Since $f_{*,a}: E \to E'$ is an isomorphism, $d(g_1 \circ f)_a|E, \ldots, d(g_q \circ f)_a|E$ are R independent, in particular, $d(g_1 \circ f)_a, \ldots, d(g_q \circ f)_a$ are independent. Since

$$f^{-1}(W) \cap f^{-1}(U') = \{x \in f^{-1}(U') | \; g_1 \circ f(x) = \dots \\ = g_q \circ f(x) = 0\},$$

it follows from corollary 2.5.5 that $f^{-1}(W)$ is a submanifold of V of codimension q.

2.16.4 LEMMA. Let K be a compact subset of V. Then the set of C^k mappings of V into V' which are transversal to W at every point of K is open in $C^k(V, V')$.

This follows easily from the definition.

Let V, V', W be now C^∞ manifolds and let K be a compact subset of V.

2.16.5 PROPOSITION. The set of C^∞ mappings of V into V' which are transversal to W at every point of K is dense in $C^\infty(V, V')$ (even in $\mathfrak{C}^\infty(V, V')$).

PROOF. Let f: $V \to V'$ be any C^∞ map. Because of lemma 2.16.4 it is sufficient to prove that any point $a \in V$ has a neighbourhood U such that the set of C^∞ maps of V into V' which are transversal to W at every point of V contains f in its closure.

Let (U_0, φ_0) be a coordinate system with $a \in U_0$, $U_0 \Subset V$ and let (U', φ') be a coordinate system on V' such that $f(U_0)$ is relatively compact in U'. Let U_1 be relatively compact in U_0, and let λ be a C^∞ function on V such that supp $(\lambda) \subset U_0$ and $\lambda(x) = 1$ for x in a neighbourhood of $\overline{U}_1$. We suppose that $0 \in \varphi'(U')$. If $u_1, \dots, u_p$ are C^k maps of V into R^m, and, if $\xi_1, \dots, \xi_p$ are real numbers, $|\xi_j| < \delta$, then for δ sufficiently small,

$$g(x; \xi_1, \dots, \xi_p) = \begin{cases} \lambda(x)(\xi_1 u_1(x) + \dots + \xi_p u_p(x)) & \text{for } x \in U_0 \\ 0 & \text{for } x \notin U_0 \end{cases}$$

defines a C^k map of V into $\varphi'(U')$, and we define a map F_ξ: $V \to V'$ by

$$F_\xi(x) = \begin{cases} \varphi'^{-1} \circ (g(x, \xi_1, \dots, \xi_p) + \varphi' \circ f(x)) & \text{for } \; x \in U_0 \\ f(x) & \text{for } \; x \notin U_0 . \end{cases}$$

This map is defined and C^∞ on V for δ sufficiently small. Let $Q = \{\xi \in R^p | \; |\xi_j| < \delta\}$. This gives us a map

$$F: Q \times V \to V', \qquad F(\xi, x) = F_\xi(x).$$

It is easily checked that if the vectors $u_1(0), \ldots, u_p(0)$ span R^m, then the map

$$F_{*,(0,a)}: T_{(0,a)}(Q \times V) \to T_b(V')$$

is surjective. Hence if δ is small enough and U is a sufficiently small neighbourhood of a, the map

$$F_{*,(\xi,a')}: T_{(\xi,a')}(Q \times V) \to T_{F(\xi,a')}(V')$$

is surjective for all $\xi \in Q$ and $a' \in U$. In particular, the map $F: Q \times U \to V$ is transversal to W. Hence (see proposition 2.16.3) $F^{-1}(W) = W_0$ is a submanifold of $Q \times U$ of codimension q (note that $W_0 \neq \emptyset$ since $F(0, a) = b \in W$).

Let π be the restriction to W_0 of the projection of $Q \times U$ onto Q. Let A_π be the critical set of π (i.e., the set of $(\xi, x) \in W_0$ such that the rank of $\pi_{*,(\xi,x)}$ is $< p = \dim Q$). Then, by Sard's theorem 2.2.13, $\pi(A_\pi)$ is nowhere dense in Q. Our proposition 2.16.5 is then an immediate consequence of the following statement:

2.16.6 PROPOSITION. For any $\xi \notin \pi(A_\pi)$, the map $F_\xi: V \to V'$ is transversal to W.

PROOF. Let $x \in V$. If $(\xi, x) \notin W_0$, then $F_\xi(x) \notin W$, so that we have nothing to prove. Hence suppose that $(\xi, x) \in W_0$. Then, since $\xi \notin \pi(A_\pi)$, the map $\pi_{*,(\xi,x)}: T_{(\xi,x)}(W_0) \to T_\xi(Q)$ is surjective. We identify $T_{(\xi,x)}(Q \times V)$ with $T_\xi(Q) \oplus T_x(V)$ and $T_{(\xi,x)}(W_0)$ with a subspace T_0 of $T_\xi(Q) \oplus T_x(V)$. Then the projection of T_0 onto the first factor $T_\xi(Q)$ is surjective. Since F is transversal to W at (ξ, x), there exists (proposition 2.16.2) a q-dimensional subspace E_0 of $T_\xi(Q) \oplus T_x(V)$ such that

$$F_{*,(\xi,x)}(E_0) = E'$$

is a q-dimensional subspace of $T_{F(\xi,x)}(V')$ with

$$E' \cap T_{F(\xi,x)}(W) = \{0\}.$$

We claim that if $\Phi = F_{*,(\xi,x)}$, we have $\Phi^{-1}(T_{F(\xi,x)}(W)) \subset T_{(\xi,x)}(W_0)$. In fact if $\Phi(v) \in T_{F(\xi,x)}(W)$, we can write $v = w_0 + v_0$, $w_0 \in T_{(\xi,x)}(W_0)$, $v_0 \in E_0$, clearly $\Phi(v_0) \in T_{F(\xi,x)}(W)$. Clearly also $\Phi(v_0) \in E'$. Hence $\Phi(v_0) = 0$. Since $\Phi|E_0$ is injective, $v_0 = 0$ and $v \in T_{(\xi,x)}(W_0)$.

It follows immediately that the projection E of E' onto $T_x(V)$ is q-dimensional and $\Phi(E)$ is a q-dimensional subspace of $T_{F(\xi,x)}(V')$ with $\Phi(E) \cap T_{F(\xi,x)}(W) = \{0\}$. Again by proposition 2.16.2, this shows that F_ξ is transversal at x.

2.16.7 THOM'S TRANSVERSALITY THEOREM. Let V, V' be C^∞ manifolds with countable base and let W be a closed submanifold of V'. Then the set of C^∞ mappings of V into V' which are transversal to W is dense in $C^\infty(V, V')$.

PROOF. V is a countable union of compact sets. Hence by lemma 2.16.4 and proposition 2.16.5, the set of C^∞ maps of V into V' transversal to W is a countable intersection of open dense sets in $C^\infty(V, V')$. Since $C^\infty(V, V')$ is a complete metric space, the theorem results from the theorem of Baire that a countable intersection of open dense sets in a complete metric space is again dense; see BOURBAKI [1958].

2.16.8 REMARK. It is actually true that the C^∞ maps transversal to W form a dense subset of $C^\infty(V, V')$ in the topology of $\mathfrak{C}^\infty(V, V')$. This follows from proposition 2.16.5 and the fact that although $\mathfrak{C}^\infty(V, V')$ is not a complete metric space, Baire's theorem is nevertheless true in this space; see CARTAN [1962]. This can also be proved along the same lines as the proof of theorem 2.15.7.

CHAPTER 3

Linear elliptic differential operators

§ 3.1 Vector bundles

3.1.1 DEFINITION. Let X, E be Hausdorff spaces and $p: E \to X$ a continuous map. The triple $\xi = (E, p, X)$ is called a *complex vector bundle* of rank q over the space X (or simply bundle over X) if the following two conditions are satisfied:

(a) For every $a \in X$, $p^{-1}(a) = E_a$ is provided with the structure of a complex vector space of dimension q.

(b) For any $a \in X$, there is a neighbourhood U of a and a homeomorphism $h = h_U$ of $p^{-1}(U) = E_U$ onto $U \times C^q$ such that if π denotes the projection of $U \times C^q$ onto U, we have: $\pi \circ h = p$ on E_U, [i.e., $\pi(h(y)) = x$ if $y \in E_x$, $x \in U$], and $h|E_x$ is a C-isomorphism of E_x onto $\{x\} \times C^q$.

$E_a = p^{-1}(a)$ is called the *fibre* of ξ at the point a. A bundle of rank 1 is called a line bundle. We can replace complex vector spaces by *real* vector spaces and so define real vector bundles in the same way.

If E, X are C^k manifolds ($1 \leqq k \leqq \infty$), if p is a C^k map and if the homeomorphisms $h_U: E_U \to U \times C^q$ can be chosen so that they are C^k diffeomorphisms, we call (E, p, X) a C^k vector bundle (or vector bundle of class C^k).

If X is a real or complex analytic manifold, real and complex analytic vector bundles can be defined in the same way. Complex analytic vector bundles are also called holomorphic vector bundles.

3.1.2 DEFINITION. If $\xi = (E, p, X)$, $\xi' = (E', p', X)$ are two complex vector bundles over X, a bundle map (or morphism) u of ξ into ξ' is a continuous map $u: E \to E'$ such that $p' \circ u = p$ and such that $u|E_a$ is a C linear map of E_a into E'_a.

We also write $u: \xi \to \xi'$ for a map $u: E \to E'$ if $p' \circ u = p$.

One can, of course, define C^k (real and complex analytic) bundle maps in the same way.

If ξ, ξ' are vector bundles, they are called isomorphic if there exist bundle maps $u: E \to E'$ and $u': E' \to E$ such that $u' \circ u$ is the identity of E, $u \circ u'$ is the identity of E'; here $\xi = (E, p, X)$, $\xi' = (E', p', X')$.

An example of a vector bundle of rank q is the following: Let $E = X \times C^q$, and $p: E \to X$ the natural projection. The triple $\vartheta_q = (E, p, X)$ is then a vector bundle $\vartheta_q = \vartheta_q(X)$ called the *trivial bundle* of rank q over X. A bundle isomorphic to ϑ_q is said to be trivial.
By definition, any vector bundle $\xi = (E, p, X)$ of rank q is locally isomorphic to the trivial bundle of rank q. Hence there is an open covering $\mathfrak{U} = \{U_i\}$ of X and homeomorphisms

$$\varphi_i: E_{U_i} = p^{-1}(U_i) \to U_i \times C^q,$$

such that if

$$U_{ij} = U_i \cap U_j \neq \emptyset,$$

then

$$\varphi_j \circ \varphi_i^{-1}: U_{ij} \times C^q \to U_{ij} \times C^q$$

is a homeomorphism and is of the form

$$\varphi_j \circ \varphi_i^{-1}(x, v) = (x, g_{ji}(x, v)),$$

where $g_{ji}(x, v)$ is, for fixed x, a C linear automorphism of C^q. Hence

$$\varphi_j \circ \varphi_i^{-1}(x, v) = (x, g_{ji}(x)v),$$

where, for $x \in U_{ij}$,

$$g_{ji}(x) \in GL(q, C).$$

It is immediately verified that for $x \in U_i \cap U_j \cap U_k$, we have

$$g_{ij}(x) g_{jk}(x) = g_{ik}(x).$$

In particular,

$$g_{ii}(x) = I$$

(the unit matrix in $GL(q, C)$) and

$$g_{ij}(x)^{-1} = g_{ji}(x).$$

Clearly the map

$$g_{ij}: x \mapsto g_{ij}(x)$$

is a continuous map of U_{ij} into $GL(q, C)$.

The g_{ij} are called *transition functions* (or transition maps) of the bundle ξ.

If ξ is a C^k (or real or complex analytic) bundle, then the g_{ij} are C^k (or real or complex analytic) mappings.

3.1.3 REMARKS. Conversely, let X be a Hausdorff topological space and $\mathfrak{U} = \{U_i\}$ an open covering of X. Let $g_{ij}: U_{ij} \to GL(q, C)$ be continuous maps so that $g_{ij}(x)g_{jk}(x) = g_{ik}(x)$ for $x \in U_i \cap U_j \cap U_k$. Let S be the topological sum $\bigcup_i(U_i \times C^q \times \{i\})$. We define an equivalence relation $\sim$ on S by

$$(x, v, i) \sim (x', v', j) \quad \text{if} \quad x' = x \quad \text{and} \quad v' = g_{ji}(x)v.$$

It is verified at once that this relation is open and that its graph is closed, so that the quotient space $E = S/\sim$ is a Hausdorff space. Let $p': S \to X$ be the map

$$(x, v, i) \mapsto x.$$

Then equivalent points have the same image under p', and so p' defines a continuous map $p: E \to X$. If $\eta: S \to E$ is the natural quotient map, we have

$$p^{-1}(U_i) = \{\eta(x, v, i) | x \in U_i, v \in C^q\},$$

and hence $\eta | U_i \times C^q \times \{i\}$ is a homeomorphism onto $p^{-1}(U_i)$. It follows at once that there is a homeomorphism

$$\varphi_i: p^{-1}(U_i) \to U_i \times C^q$$

having the properties required in definition 3.1.1. Thus $\xi = (E, p, X)$ is a vector bundle. It follows immediately that the transition functions of ξ (relative to the covering $\mathfrak{U}$) are precisely the maps g_{ij}.

If X is a C^k (real, complex analytic) manifold and the g_{ij} are C^k (real, complex analytic) maps, then the bundle constructed above becomes a C^k (real, complex analytic) bundle.

3.1.4 Remark. Let $\xi = (E, p, X)$ be a vector bundle and, with respect to an open covering $\mathfrak{U} = \{U_i\}$ of X, let

$$g_{ij}: U_i \cap U_j \to GL(q, \mathbf{C})$$

be transition maps of ξ. Let $\{V_\alpha\}_{\alpha \in \mathscr{A}}$ be a refinement of $\mathfrak{U}$, and let

$$\tau: \mathscr{A} \to \mathscr{I}$$

be a 'refinement map', i.e. a map such that $V_\alpha \subset U_{\tau(\alpha)}$, $\alpha \in \mathscr{A}$. Let $h_{\alpha\beta}$ be the restriction to $V_\alpha \cap V_\beta$ of $g_{\tau(\alpha)\,\tau(\beta)}$; let $\xi' = (E', p', X)$ be the bundle constructed as in remark 3.1.3 with the 'transition functions' $\{h_{\alpha\beta}\}$. Then E' is a quotient of

$$S' = \bigcup_{\alpha \in \mathscr{A}} (V_\alpha \times \mathbf{C}^q \times \{\alpha\}).$$

Let $u': S' \to E$ be the map defined by

$$u'(x, v, \alpha) = \varphi_i^{-1}(x, v),$$

where $i = \tau(\alpha)$ and $\varphi_i: p^{-1}(U_i) \to U_i \times \mathbf{C}^q$ are isomorphisms inducing the transition functions g_{ij}. Then u' induces an isomorphism $u: \xi \to \xi'$.

3.1.5 Proposition. Let $\xi = (E, p, X)$, $\xi' = (E', p', X)$ be two vector bundles and $\mathfrak{U} = \{U_i\}$, $\mathfrak{U}' = \{U'_\alpha\}$ two open coverings of X. Let $\{g_{ij}\}$, $\{g'_{\alpha\beta}\}$ be transition maps for ξ, ξ' respectively relative to the coverings $\mathfrak{U}$, respectively $\mathfrak{U}'$. Then ξ and ξ' are isomorphic if and only if the following condition is satisfied:

There exists a common refinement

$$\mathfrak{B} = \{V_\lambda\}_{\lambda \in \Lambda}$$

of $\mathfrak{U}$ and $\mathfrak{U}'$, and refinement maps

$$\tau: \Lambda \to \mathscr{I}, \qquad \tau': \Lambda \to \mathscr{A}$$

such that

$$V_\lambda \subset U_{\tau(\lambda)} \cap U'_{\tau'(\lambda)};$$

further there exist continuous maps $h_\lambda: V_\lambda \to GL(q, C)$ with the property that

$$h_\lambda(x) g_{\tau(\lambda)\tau(\mu)}(x) h_\mu(x)^{-1} = g'_{\tau'(\lambda)\tau'(\mu)}(x)$$

for $x \in V_\lambda \cap V_\mu$.

PROOF. By remark 3.1.4, we can assume that $\mathfrak{U} = \mathfrak{U}'$, and that they are indexed by the same set Λ. We denote this covering by $\mathfrak{B}$.

Suppose that $u: \xi \to \xi'$ is an isomorphism. Let

$$\varphi_\lambda: p^{-1}(V_\lambda) \to V_\lambda \times C^q, \qquad \varphi'_\lambda: p'^{-1}(V_\lambda) \to V_\lambda \times C^q$$

be isomorphisms inducing the transition functions $g_{\lambda\mu}$, $g'_{\lambda\mu}$ respectively. Then

$$u_\lambda = \varphi'_\lambda \circ u \circ \varphi_\lambda^{-1}$$

is an isomorphism of $V_\lambda \times C^q$ onto itself, so that there exist continuous maps

$$h_\lambda: V_\lambda \to GL(q, C)$$

with

$$u_\lambda(x, v) = (x, h_\lambda(x)v), \qquad x \in V_\lambda, \quad v \in C^q.$$

Then

$$u_\lambda \circ \varphi_\lambda \circ \varphi_\mu^{-1} \circ u_\mu^{-1} = \varphi'_\lambda \circ \varphi'^{-1}_\mu.$$

Since $\varphi_\lambda \circ \varphi_\mu^{-1}(x, v) = (x, g_{\lambda\mu}(x)v)$ and a similar formula holds for $\varphi_\lambda' \circ \varphi'^{-1}_\mu$, we obtain

$$h_\lambda g_{\lambda\mu} h_\mu^{-1} = g'_{\lambda\mu}.$$

Conversely, suppose given continuous maps

$$h_\lambda: V_\lambda \to GL(q, C)$$

with

$$h_\lambda g_{\lambda\mu} h_\mu^{-1} = g'_{\lambda\mu}$$

and let $\varphi_\lambda: p^{-1}(V_\lambda) \to V_\lambda \times C^q$, $\varphi'_\lambda: p'^{-1}(V_\lambda) \to V_\lambda \times C^q$ be as above. Let

$$u_\lambda(x, v) = (x, h_\lambda(x)v),$$

and consider the map

$$\varphi'^{-1}_\lambda \circ u_\lambda \circ \varphi_\lambda: p^{-1}(V_\lambda) \to p'^{-1}(V_\lambda).$$

Clearly this is an isomorphism. Furthermore it is immediately verified that on $p^{-1}(V_\lambda) \cap p^{-1}(V_\mu)$, we have

$$\varphi_\lambda'^{-1} \circ u_\lambda \circ \varphi_\lambda = \varphi_\mu'^{-1} \circ u_\mu \circ \varphi_\mu .$$

This gives us an isomorphism $u: \xi \to \xi'$.

3.1.6 REMARK. Let $\xi_1 = (E_1, p_1, X)$, $\xi_2 = (E_2, p_2, X)$ be two vector bundles. Let $a \in X$ and put $E_a = E_{1,a} \oplus E_{2,a}$. Let E be the disjoint union

$$E = \bigcup_{a \in X} E_a$$

and let $p: E \to X$ be the map $e \mapsto a$ if $e \in E_a$.

If U is an open set in X such that there are isomorphisms

$$\varphi_1 : p_1^{-1}(U) \to U \times C^{q_1}, \qquad \varphi_2 : p_2^{-1}(U) \to U \times C^{q_2},$$

we have a *bijection*

$$\varphi : p^{-1}(U) \to U \times C^{q_1 + q_2}$$

defined by $e_1 \oplus e_2 \mapsto (x, v_1, v_2)$ where $\varphi_1(e_1) = (x, v_1)$, $\varphi_2(e_2) = (x, v_2)$. It is immediately clear that there is a unique topology on E such that the sets $p^{-1}(U)$ are open in E and all the maps φ defined as above are homeomorphisms. With this topology, $\xi = (E, p, X)$ is a vector bundle called the *direct sum* (or *Whitney sum*) of ξ_1 and ξ_2. We write

$$\xi = \xi_1 \oplus \xi_2 .$$

If, relative to an open covering $\{U_i\}$ of X, ξ_1, ξ_2, are given by transition maps $g_{ij}^{(1)}$, $g_{ij}^{(2)}$, then the transition maps of ξ obtained from the above isomorphisms are given by

$$g_{ij}(x) = g_{ij}^{(1)}(x) \oplus g_{ij}^{(2)}(x) = \begin{pmatrix} g_{ij}^{(1)}(x) & 0 \\ 0 & g_{ij}^{(2)}(x) \end{pmatrix} .$$

3.1.7 REMARK. If ξ_1, ξ_2 are vector bundles over X as above, one can define, in the same way, vector bundles

$$\xi_1 \otimes \xi_2, \qquad \mathrm{Hom}\,(\xi_1, \xi_2) \text{ and } \lambda^p(\xi_1)$$

such that, for any $a \in X$, we have

$$E_a = E_{1,a} \otimes E_{2,a}, \qquad E_a = \mathrm{Hom}_C(E_{1,a}, E_{2,a}), \qquad E_a = \wedge^p E_{1,a}$$

respectively, where (E, p, X) stands for the corresponding bundle. If $\xi_2 = \vartheta_1$ is the trivial bundle of rank 1, the bundle

$$\mathrm{Hom}(\xi_1, \vartheta_1) = \xi_1^*$$

is called the *dual* of ξ_1; its fibre at $a \in X$ is the dual of the fibre of ξ_1 at a. The vector bundle $\xi_1 \otimes \xi_2$ is called the *tensor product* of the bundles ξ_1, ξ_2, and $\lambda^p(\xi_1)$ is called the *pth exterior power of* ξ_1.

If $\{U_i\}$ is an open covering of X and ξ_1, ξ_2 give rise to transition maps $g_{ij}^{(1)}, g_{ij}^{(2)}$ respectively, then $\xi = \xi_1 \otimes \xi_2$ has transition maps g_{ij} defined by

$$g_{ij}(x) = g_{ij}^{(1)}(x) \otimes g_{ij}^{(2)}(x),$$

where the latter term denotes the Kronecker (or tensor) product of the matrices $g_{ij}^{(1)}(x)$ and $g_{ij}^{(2)}(x)$. Further ξ_1^* has transition maps g_{ij}^* defined by

$$g_{ij}^*(x) = {}^t g_{ij}(x)^{-1},$$

where ${}^tA$ denotes the transpose of the matrix A.

The isomorphism

$$L \otimes M^* \simeq \mathrm{Hom}(M, L)$$

when L, M are finite dimensional vector spaces and M^* is the dual of M, gives rise to an isomorphism

$$\xi_1 \otimes \xi_2^* \simeq \mathrm{Hom}(\xi_2, \xi_1)$$

if ξ_1, ξ_2 are two vector bundles. In particular, for any vector bundle ξ, there is a natural isomorphism

$$\xi \otimes \xi^* \simeq \mathrm{Hom}(\xi, \xi).$$

Let $\xi_1, \xi_2, \xi_1', \xi_2'$ be vector bundles on X and let

$$u_k: \xi_k \to \xi_k', \qquad k = 1, 2$$

be bundle maps. For every $a \in X$, we have C linear maps

$$u_{k,a}: E_{k,a} \to E_{k,a}',$$

$\{\xi_k = (E_k, p_k, X), \xi'_k = (E'_k, p'_k, X)\}$. This gives us $\mathbf{C}$ linear maps

$$u_{1,a} \otimes u_{2,a}: E_{1,a} \otimes E_{2,a} \to E'_{1,a} \otimes E'_{2,a},$$

which in turn define a bundle map

$$u_1 \otimes u_2: \xi_1 \otimes \xi_2 \to \xi'_1 \otimes \xi'_2.$$

In the same way, for any bundle ξ, the bundle map $u_1: \xi_1 \to \xi'_1$ induces bundle maps

$$\operatorname{Hom}(\xi, \xi_1) \to \operatorname{Hom}(\xi, \xi'_1); \qquad \operatorname{Hom}(\xi'_1, \xi) \to \operatorname{Hom}(\xi_1, \xi).$$

In particular, we have the bundle maps

$$u_1^*: \xi_1'^* \to \xi_1^*, \qquad \lambda^p(u_1): \lambda^p(\xi_1) \to \lambda^p(\xi'_1).$$

3.1.8 REMARK. We remark that corresponding definitions can be given for real vector bundles as well as for C^k and real and complex analytic vector bundles.

3.1.9 EXAMPLE. If V is a C^k (real analytic) manifold of dimension n, then the tangent bundle

$$T(V) = \bigcup_{a \in V} T_a(V)$$

introduced in § 2.2 is a real C^{k-1} (real analytic) bundle of rank n; so is the cotangent bundle

$$T^*(V) = \bigcup_{a \in V} T_a^*(V).$$

The bundle of p-forms $\wedge^p T^*(V)$ introduced in theorem 2.2.8 is nothing but the pth exterior power $\lambda^p(T^*(V))$ of $T^*(V)$.

If V is a complex analytic manifold, all these bundles are holomorphic vector bundles. Moreover, on a complex manifold V, we can define the bundle of forms of type (p, q):

$$\mathscr{E}^*_{p,q}(V) = \bigcup_{a \in V} \mathscr{E}^*_{p,q}(V, a) \quad \text{(see remark 2.4.10).}$$

This is a real analytic complex vector bundle on V.

3.1.10 DEFINITION. Let V be a C^k manifold and $\xi = (E, p, X)$ a C^k vector bundle on V. Let U be open in V. Then a C^k section s of ξ over U is a C^k map $s\colon U \to E$ such that $p \circ s =$ identity on U. The set of these sections will be denoted by $C^k(U, \xi)$.

We will also have occasion to consider not necessarily continuous sections of ξ; these are simply set mappings $s\colon U \to E$ with $p \circ s =$ identity.

If s is an arbitrary section of ξ over U, the support of s (in U) is the closure in U of the set $\{a \in U \mid s(a) \neq 0_a\}$; here 0_a denotes the zero element of the vector space $E_a = p^{-1}(a)$. We shall usually drop the suffix a and write simply 0 for 0_a.

The set of C^k sections over U having compact support is denoted by $C_0^k(U, \xi)$.

Note that if $\xi = \vartheta_q$ is the trivial bundle of rank q, the set $C^k(V, \vartheta_q)$ can be canonically identified with the set of q-tuples of C^k functions on U. As in ch. 1, we write $C^k(U, q)$ for this set. Similarly, $C_0^k(U, \vartheta_q) = C_0^k(U, q)$ is the set of q-tuples of C^k functions with compact support. Let $\xi = (E, p, X)$ be a vector bundle having transition maps g_{ij} with respect to an open covering $\{U_i\}$, and

$$\varphi_i\colon p^{-1}(U_i) \to U_i \times C^q$$

corresponding isomorphisms. If s is a section of ξ over X, the maps

$$s_i = \varphi_i \circ s\colon U_i \to U_i \times C^q$$

define maps

$$\sigma_i\colon U_i \to C^q.$$

Since $\varphi_i^{-1} \circ s_i = s$, we see that

$$s_i = \varphi_i \circ \varphi_j^{-1} \circ s_j,$$

which gives

$$\sigma_i(x) = g_{ij}(x)\sigma_j(x) \qquad \text{for} \quad x \in U_i \cap U_j.$$

Conversely, maps $\sigma_i\colon U_i \to C^q$ with $\sigma_i(x) = g_{ij}(x)s_j(x)$ for $x \in U_i \cap U_j$ define a section s of ξ by

$$s = \varphi_i^{-1} \circ s_i \quad \text{on} \quad U_i, \qquad s_i(x) = (x, \sigma_i(x)).$$

This section is C^k (real or complex analytic) if and only if the maps σ_i are.

3.1.11 PROPOSITION. If ξ is a bundle of rank 1 (i.e. a line bundle) then ξ is isomorphic to ϑ_1 (i.e. ξ is trivial) if and only if ξ has a continuous section s such that $s(a) \neq 0$ for each a.

PROOF. To see this, we have only to prove that if there is such a section s, then ξ is trivial. Now, we obtain a map $\xi \to \vartheta_1$ as follows: if $e \in p^{-1}(a)$, then there is a unique $\lambda \in C$ with $e = \lambda s(a)$. The map

$$e \mapsto (a, \lambda)$$

is an isomorphism of ξ onto ϑ_1.
One sees from this that if ξ is a line bundle, then $\xi \otimes \xi^*$ is trivial. In fact

$$\xi \otimes \xi^* \simeq \mathrm{Hom}\,(\xi, \xi)$$

and the section s of Hom (ξ, ξ), defined by

$$s(a) = \text{identity of } p^{-1}(a),$$

is nowhere zero.

Of course, corresponding remarks apply to C^k and to real and complex analytic line bundles.

3.1.12 REMARK. If $\xi = (E, p, V)$ is a vector bundle, then, for any open set $U \subset V$, the triple $(p^{-1}(U), p, U)$ is again a vector bundle. We denote this by $\xi | U$. We use similar notation also for a closed subset of V.

§ 3.2 Fourier transforms

Let Ω be an open set in R^n and p a real number $\geqq 1$. The set of complex-valued Lebesgue measurable functions f on Ω such that $|f|^p$ is integrable with respect to Lebesgue measure forms a vector space. The quotient of this space by the subspace of functions zero almost everywhere is a Banach space relative to the norm

$$\|f\|_{L^p} = \left(\int_\Omega |f(x)|^p \mathrm{d}x\right)^{1/p}.$$

We denote this Banach space by $L^p = L^p(\Omega)$. We shall identify elements of this space with representative functions when no confusion is likely. If $p = 2$ and $f, g \in L^2(\Omega)$, we set

$$(f, g)_{L^2} = \int_\Omega f(x)\overline{g(x)}\mathrm{d}x.$$

$L^2(\Omega)$ is a Hilbert space relative to this scalar product.

Let $f \in L^1(R^n)$. The Fourier transform $\hat{f}$ of f is defined by

$$\hat{f}(\xi) = (2\pi)^{-n/2}\int_{R^n} f(x)\mathrm{e}^{-\mathrm{i}\langle x, \xi\rangle}\mathrm{d}x,$$

where $\xi = (\xi_1, \ldots, \xi_n)$ and $\langle x, \xi\rangle = x_1\xi_1 + \ldots + x_n\xi_n$.

Let $\mathscr{S}$ be the set of C^∞ functions f on R^n such that for any integer $N \geqq 0$ and any $\alpha = (\alpha_1, \ldots, \alpha_n)$,

$$(1+|x|^2)^N D^\alpha f(x)$$

is bounded on R^n. $\mathscr{S}$ is called the Schwartz space or space of rapidly decreasing functions on R^n. For any real number p, $1 \leqq p < \infty$, $\mathscr{S}$ is contained in L^p as a dense subspace. Moreover, for any $\alpha = (\alpha_1, \ldots, \alpha_n)$ and $f \in \mathscr{S}$, $D^\alpha f$ also lies in $\mathscr{S}$. In particular, all derivatives of f are bounded. For $f \in \mathscr{S}$, we have

$$(D^\alpha f)\hat{\ }(\xi) = \mathrm{i}^{|\alpha|}\xi^\alpha \hat{f}(\xi)$$

and

$$(D^\alpha \hat{f})(\xi) = ((-\mathrm{i}x)^\alpha f(x))\hat{\ }.$$

Further, if $f \in L^1(R^n)$, then for every $\xi \in R^n$, we have

$$|\hat{f}(\xi)| \leqq \|f\|_{L^1}.$$

From these remarks, we deduce:

3.2.1 Corollary. If $f \in \mathscr{S}$, then $\hat{f} \in \mathscr{S}$.
In what follows, all integrals in which the domain of integration is not explicitly given are taken over R^n.

3.2.2 The inversion formula. If $f \in \mathscr{S}$, we have

$$f(x) = (2\pi)^{-n/2}\int_{R^n} \hat{f}(\xi)\mathrm{e}^{\mathrm{i}\langle x, \xi\rangle}\mathrm{d}\xi, \qquad x \in R^n.$$

PROOF. Let $\varphi \in \mathscr{S}$. Then, since $\hat{f}$ is bounded, the function $\varphi(\xi)\hat{f}(\xi)$ is integrable. By Fubini's theorem, we obtain:

3.2.3

$$\int \varphi(\xi)\hat{f}(\xi)e^{i\langle x,\xi\rangle}d\xi = (2\pi)^{-n/2}\int \varphi(\xi)e^{i\langle x,\xi\rangle}\left\{\int f(y)e^{-i\langle y,\xi\rangle}dy\right\}d\xi$$

$$= (2\pi)^{-n/2}\int f(y)dy\int \varphi(\xi)e^{-i\langle y-x,\xi\rangle}d\xi$$

$$= (2\pi)^{-n/2}\int f(x+t)dt\int \varphi(\xi)e^{-i\langle t,\xi\rangle}d\xi$$

$$= \int f(x+t)\hat{\varphi}(t)dt.$$

We now choose $g \in \mathscr{S}$ and set $\varphi(\xi) = g(\varepsilon\xi)$ with $\varepsilon > 0$. Then

$$\hat{\varphi}(t) = \varepsilon^{-n}\hat{g}(t/\varepsilon).$$

This gives

$$\int g(\varepsilon\xi)\hat{f}(\xi)e^{i\langle x,\xi\rangle}d\xi = \int f(x+t)\varepsilon^{-n}\hat{g}(t/\varepsilon)dt.$$

$$= \int f(x+\varepsilon t)\hat{g}(t)dt.$$

Since f and g are bounded and $\hat{f}$ and $\hat{g}$ are integrable (being in $\mathscr{S}$ by corollary 3.2.1), we can take limits as $\varepsilon \to 0$ under the integral, and we get

$$g(0)\int \hat{f}(\xi)e^{i\langle x,\xi\rangle}d\xi = f(x)\int \hat{g}(t)dt.$$

If we take

$$g(\xi) = \exp\left(-\tfrac{1}{2}(\xi_1^2 + \ldots + \xi_n^2)\right),$$

we see at once that

$$\hat{g}(t) = g(t)$$

so that

$$g(0) = (2\pi)^{-n/2}\int \hat{g}(t)dt = 1.$$

Using this relation in our formula above, we get

$$f(x) = (2\pi)^{-n/2}\int \hat{f}(\xi)e^{i\langle x,\xi\rangle}d\xi.$$

3.2.4 COROLLARY. If f, $\varphi \in \mathscr{S}$, we have

$$\int f(\xi)\hat{\varphi}(\xi)\mathrm{d}\xi = \int \varphi(\xi)\hat{f}(\xi)\mathrm{d}\xi.$$

This follows from (3.2.3) if we put $x = 0$.

3.2.5 COROLLARY. If $f \in L^1(R^n)$ and $g \in \mathscr{S}$, we have

$$\int g(\xi)\hat{f}(\xi)\mathrm{e}^{\mathrm{i}\langle x, \xi\rangle}\mathrm{d}\xi = \int f(x+t)\hat{g}(t)\mathrm{d}t.$$

This follows from the proof of (3.2.3).

3.2.6 COROLLARY. If $f, g \in \mathscr{S}$, we have

$$(f, g)_{L^2} = (\hat{f}, \hat{g})_{L^2}.$$

In particular, for any $f \in \mathscr{S}$, we have $\|f\|_{L^2} = \|\hat{f}\|_{L^2}$.

PROOF. Let $\varphi \in \mathscr{S}$ be defined by

$$\overline{\varphi(t)} = \hat{g}(t).$$

Now, by formula 3.2.2, we have

$$g(x) = (2\pi)^{-n/2}\int \hat{g}(\xi)\mathrm{e}^{\mathrm{i}\langle x, \xi\rangle}\mathrm{d}\xi.$$

Taking conjugates, we see that

$$\overline{g(x)} = (2\pi)^{-n/2}\int \varphi(\xi)\mathrm{e}^{-\mathrm{i}\langle x, \xi\rangle}\mathrm{d}\xi = \hat{\varphi}(x).$$

The corollary follows at once from corollary 3.2.4.

3.2.7 REMARK. If $f \in L^p(R^n)$ for some $p \geqq 1$, then for any $g \in \mathscr{S}$, the product fg is integrable. For $f \in L^p(R^n)$, we define a linear functional $\hat{f}$ on $\mathscr{S}$ by

$$(\hat{f}, \varphi) = \int f(t)\hat{\varphi}(t)\mathrm{d}t, \qquad \varphi \in \mathscr{S}.$$

We call $\hat{f}$, again, the Fourier transform of f. If, for some q, $1 \leqq q \leqq \infty$, there is $g \in L^q(R^n)[L^\infty(R^n)$ is the set of measurable functions h, which are bounded outside a set of measure zero] such that

$$(\hat{f}, \varphi) = \int g(x)\varphi(x)\mathrm{d}x, \qquad \varphi \in \mathscr{S},$$

we say that $\hat{f}$ belongs to $L^q(R^n)$ and we identify $\hat{f}$ with g and write $\hat{f} = g$. Note that g, if it exists, is uniquely determined upto values on a set of measure zero. Furthermore, when $f \in L^1(R^n)$, this definition is compatible with the earlier one in view of corollary 3.2.5.

3.2.8 PLANCHEREL'S THEOREM. If $f \in L^2(R^n)$, then $\hat{f} \in L^2(R^n)$ and $\|f\|_{L^2} = \|\hat{f}\|_{L^2}$.

PROOF. Let $f \in L^2(R^n)$. There is a sequence $\{f_\nu\}$ of elements in $\mathscr{S}$ such that $\|f - f_\nu\|_{L^2} \to \infty$. Now, by corollary 3.2.6,

$$\|\hat{f}_\nu - \hat{f}_\mu\|_{L^2} = \|f_\nu - f_\mu\|_{L^2},$$

and it follows that

$$\|\hat{f}_\nu - \hat{f}_\mu\|_{L^2} \to 0,$$

as $\nu, \mu \to \infty$. Because of the completeness of $L^2(R^n)$, there is $g \in L^2(R^n)$ so that

$$\|g - \hat{f}_\nu\|_{L^2} \to 0.$$

Clearly

$$\|g\|_{L^2} = \lim_{\nu \to \infty} \|\hat{f}_\nu\|_{L^2} = \lim_{\nu \to \infty} \|f_\nu\|_{L^2} = \|f\|_{L^2}.$$

Moreover, for $\varphi \in \mathscr{S}$, we have

$$\int g(t)\varphi(t)\mathrm{d}t = \lim_{\nu \to \infty} \int \hat{f}_\nu(t)\varphi(t)\mathrm{d}t = \lim_{\nu \to \infty} \int f_\nu(t)\hat{\varphi}(t)\mathrm{d}t$$

by corollary 3.2.4. Since $f_\nu \to f$ in L^2 and $\hat{\varphi} \in L^2$, it follows from Schwarz' inequality that

$$\int g(t)\varphi(t)\mathrm{d}t = \int f(t)\hat{\varphi}(t)\mathrm{d}t = (\hat{f}, \varphi)$$

for all $\varphi \in \mathscr{S}$.

This proves the theorem.

3.2.9 PROPOSITION. The inversion formula 3.2.2 can be written

$$\hat{\hat{f}}(-x) = f(x) \qquad \text{for } f \in \mathscr{S}.$$

It is an immediate consequence of Plancherel's theorem that this relation holds also for any $f \in L^2$ (equality means, of course, equality almost everywhere).

Now, if $f \in L^1(R^n)$, we have, for $\varphi \in \mathscr{S}$,

$$\int \varphi(\varepsilon\xi)\hat{f}(\xi)e^{i\langle x,\xi\rangle}\,d\xi = \int f(x+\varepsilon t)\hat{\varphi}(t)dt$$

(by corollary 3.2.5). If, in addition, $\hat{f} \in L^1$, then the term on the left converges to

$$\varphi(0)\int \hat{f}(\xi)e^{i\langle x,\xi\rangle}d\xi,$$

for any $x \in R^n$. Moreover, since $f \in L^1$, the term on the right converges, in the space $L^1(R^n)$, to

$$f(x)\int \hat{\varphi}(t)dt.$$

It follows from this that *if f and $\hat{f} \in L^1(R^n)$, then*

$$\hat{\hat{f}}(-x) = f(x)$$

almost everywhere. In particular, f is equal (almost everywhere), to a bounded continuous function.

3.2.10 PROPOSITION. If $f, g \in L^1(R^n)$, then

$$\int |f(x-y)g(y)|dy < \infty,$$

for almost all x. Further, if we set

$$(f * g)(x) = \int f(x-y)g(y)dy,$$

we have

$$\|f * g\|_{L^1} \leqq \|f\|_{L^1}\|g\|_{L^1}.$$

PROOF. It is sufficient to prove this when $f \geqq 0$, $g \geqq 0$. Now, by Fubini's theorem

$$\int dx \int f(x-y)g(y)dy = \int g(y)dy \int f(x-y)dx$$
$$= \left(\int g(y)dy\right)\left(\int f(x)dx\right) < \infty$$

and the proposition follows.

3.2.11 REMARK. If $f, g \in L^2(R^n)$, then

$$\int |f(x-y)g(y)|dy < \infty,$$

for almost all x and

$$\left|\int f(x-y)g(y)dy\right| \leqq \|f\|_{L^2}\|g\|_{L^2}.$$

This is a trivial consequence of the Schwarz inequality.

3.2.12 PROPOSITION. If $f, g \in \mathscr{S}$, then $f * g \in \mathscr{S}$ and we have

$$(f * g)\hat{\ } = (2\pi)^{n/2}\hat{f}\hat{g}.$$

PROOF. We have

$$|x|^2 = |x-y+y|^2 \leqq 2|x-y|^2 + 2|y|^2,$$

so that

$$1+|x|^2 \leqq 2(1+|x-y|^2)(1+|y|^2).$$

Hence, by remark 3.2.11, for any $\alpha = (\alpha_1, \ldots, \alpha_n)$, and $x \in R^n$, we have

$$(1+|x|^2)^N|D^\alpha(f * g)(x)| \leqq 2^N\|f'\|_{L^2}\|g'\|_{L^2},$$

where

$$f'(x) = (1+|x|^2)^N D^\alpha f(x), \qquad g'(x) = (1+|x|^2)^N g(x);$$

hence $f * g \in \mathscr{S}$. Now

$$\begin{aligned}(f*g)\hat{}(\xi) &= (2\pi)^{-n/2}\int e^{-i\langle x,\xi\rangle}\,dx\int f(x-y)g(y)dy\\ &= (2\pi)^{-n/2}\int g(y)dy\int f(x-y)e^{-i\langle x,\xi\rangle}dx\\ &= (2\pi)^{-n/2}\int g(y)e^{-i\langle y,\xi\rangle}dy\int f(x)e^{-i\langle x,\xi\rangle}dx\\ &= (2\pi)^{n/2}\hat{f}(\xi)\hat{g}(\xi).\end{aligned}$$

3.2.13 Corollary. If $f, g \in \mathscr{S}$, we have

$$(fg)\hat{} = (2\pi)^{-n/2}\hat{f}*\hat{g}.$$

This follows at once from proposition 3.2.12 and formula 3.2.2.

3.2.14 Remark. If $f \in L^1(R^n)$ or $f \in L^2(R^n)$ and $g \in \mathscr{S}$, then again $f*g$ is defined (by proposition 3.2.10 and remark 3.2.11). Furthermore, if $f_\nu \in \mathscr{S}$ and $f_\nu \to f$ in L^1 or L^2 according as $f \in L^1$ or L^2, then $\hat{f}_\nu * \hat{g}$ converges uniformly to $\hat{f}*\hat{g}$ (by remark 3.2.11 if $f \in L^2$, and because $|\hat{f}_\nu(\xi)-\hat{f}(\xi)| \leqq \|f_\nu - f\|_{L^1} \to 0$ if $f \in L^1$). Also, it is seen easily that in either case $\hat{f}_\nu\hat{g}$ converges to $\hat{f}\hat{g}$ in L^1. It follows that we have

$$(f*g)\hat{} = (2\pi)^{n/2}\hat{f}\hat{g}$$

if $g \in \mathscr{S}$ and $f \in L^1(R^n)$ or $f \in L^2(R^n)$. One sees in the same way that under these hypotheses,

$$(fg)\hat{} = (2\pi)^{-n/2}\hat{f}*\hat{g}.$$

3.2.15 Remark. All these results extend at once to functions with values in a finite dimensional C vector space. We shall use them in this more general case without explicit mention.

§ 3.3 Linear differential operators.

Let V be a C^∞ manifold of dimension n and $\xi = (E, p, V)$ and $\eta = (F, q, V)$ be two C^∞ vector bundles over V. We assume that rank $\xi = r$, rank $\eta = s$.

3.3.1 DEFINITION. A *linear differential operator* (or differential operator) P from ξ to η is a C linear map

$$P: C^\infty(V, \xi) \to C^\infty(V, \eta)$$

such that

$$\operatorname{supp}(Ps) \subset \operatorname{supp}(s)$$

for any $s \in C^\infty(V, \xi)$.

Here $C^\infty(V, \xi)$, $C^\infty(V, \eta)$ denote respectively the space of C^∞ sections of ξ and η.

Note that this gives rise at once to a C linear map

$$P_U: C^\infty(U, \xi) \to C^\infty(U, \eta),$$

for any open set $U \subset V$ which is also a linear differential operator; in fact, if $a \in U$ let φ be a C^∞ function with compact support in U which equals 1 in a neighbourhood of a. For every $s \in C^\infty(U, \xi)$, define a section $\varphi s \in C^\infty(V, \xi)$ by

$$(\varphi s)(x) = \begin{cases} \varphi(x)s(x) & \text{if } x \in U, \\ 0 & \text{if } x \notin U. \end{cases}$$

We may set

$$(P_U s)(a) = P(\varphi s)(a).$$

3.3.2 PROPOSITION. Let Ω be an open set in R^n and let P be a linear differential operator from ϑ_r to ϑ_s. Then, for any $a \in \Omega$, there exists a neighbourhood U of a, an integer $m > 0$ and a constant $C > 0$ such that

$$||Pu||_0 \leqq C||u||_m$$

for any

$$u \in C_0^\infty(U - \{a\}, r).$$

We recall that the norms on $C_0^\infty(U, r)$ are defined by

$$||u||_m = \sum_{|\alpha| \leqq m} \sum_{j=1}^{r} \frac{1}{\alpha!} \sup_{x \in U} |D^\alpha u_j(x)|, \qquad u = (u_1, \ldots, u_r).$$

PROOF OF (3.3.2). Suppose that the lemma is false in the neighbourhood of $a \in \Omega$. Let U_0 be relatively compact in Ω. Then there is an open set

$$U_1 \Subset U_0 - \{a\}$$

and an

$$u_1 \in C_0^\infty(U_1, r),$$

such that

$$||Pu_1||_0 > 2^2 ||u_1||_1 .$$

Now, $U_0 - \overline{U}_1$ is a neighbourhood of a, and, by our assumption, there is an open set

$$U_2 \Subset U_0 - \overline{U}_1 - \{a\}$$

and an

$$u_2 \in C_0^\infty(U_2, r)$$

so that

$$||Pu_2||_0 > 2^{2.2} ||u_2||_2 .$$

By induction, we can construct a sequence $\{U_k\}$ of open sets with

$$\overline{U}_k \subset U_0 - \{a\}, \qquad \overline{U}_k \cap \overline{U}_l = \emptyset \qquad \text{if} \quad k \neq l$$

and

$$u_k \in C_0^\infty(U_k, r),$$

such that

$$||Pu_k||_0 > 2^{2.k} ||u_k||_k .$$

Let

$$u = \sum_{k=1}^{\infty} \frac{2^{-k} u_k}{||u_k||_k} .$$

Clearly, the series converges in C^∞ and so

$$u \in C_0^\infty(U', r)$$

if U' is a relatively compact neighbourhood of $\overline{U}_0$ in Ω. Furthermore

$$u | U_k = 2^{-k} ||u_k||_k^{-1} u_k | U_k .$$

From the fact that supp $(Pf) \subset$ supp (f) for all f, it follows that

$$Pu | U_k = 2^{-k} ||u_k||_k^{-1} (Pu_k) | U_k .$$

Since $||Pu_k||_0 > 2^{2k} ||u_k||_k$, there exists $x_k \in U_k$ such that

$$|Pu_k(x_k)| > 2^{2k} ||u_k||_k ,$$

so that $|Pu(x_k)| > 2^k$. On the other hand, since $u \in C^\infty(\Omega, r)$, Pu is continuous on Ω and so is bounded on $\bar{U}_0$. This is a contradiction, and so proposition 3.3.2 is established.

3.3.3 THEOREM OF PEETRE. Let Ω be an open set in R^n and P a linear differential operator from ϑ_r to ϑ_s. Then, for any relatively compact subset $\Omega' \Subset \Omega$, there is an $m \geqq 0$ and there are C^∞ maps a_α of Ω' into the space of linear maps of R^r to R^s (that is to say $r \times s$ matrices) such that for any $u \in C^\infty(\Omega', r)$ and $x \in \Omega'$ we have

3.3.4 $$(Pu)(x) = \sum_{|\alpha| \leqq m} a_\alpha(x)(D^\alpha u)(x).$$

PROOF. Let U be any open subset of Ω, and *assume* that there exist constants $C > 0$, $m > 0$ such that

3.3.5 $$\|Pu\|_0 \leqq C\|u\|_m \quad \text{for all} \quad u \in C_0^\infty(U, r).$$

We first prove:

3.3.6. If $u \in C_0^\infty(U, r)$ and u is m-flat at $a \in U$ (definition 1.5.1), then $(P\,u)(a) = 0$.

By lemma 1.5.2 there exists a sequence $\{u_\nu\}$ of elements in $C_0^\infty(U, r)$, which vanish in the neighbourhood of a, and such that $\|u_\nu - u\|_m$ tends to zero as $\nu \to \infty$. By (3.3.5), Pu_ν converges uniformly to Pu on U. Now, since

$$\operatorname{supp}(Pu_\nu) \subset \operatorname{supp}(u_\nu)$$

and u_ν is zero near a, it follows that

$$(Pu_\nu)(a) = 0.$$

Hence

$$(Pu)(a) = \lim (Pu_\nu)(a) = 0.$$

Let $e_1, \ldots, e_r$ be a basis of R^r. If $u \in C^\infty(U, r)$, we have

$$u = \sum_{j=1}^{r} u_j e_j, \qquad u_j \in C^\infty(U, 1).$$

For $a \in U$, let $\mu_{\alpha, a}$ be the monomial

$$\mu_{\alpha, a}(x) = (x - a)^\alpha.$$

Then we have

$$u = \sum_{|\alpha| \leqq m} \frac{1}{\alpha!} \mu_{\alpha,a} \sum_{j=1}^{r} D^\alpha u_j e_j + f,$$

where f is m-flat at a. Hence, by (3.3.6), we have

$$(Pu)(a) = \sum_{|\alpha| \leqq m} \sum_{j=1}^{r} \frac{1}{\alpha!} P(\mu_{\alpha,a} e_j)(a) D^\alpha u_j(a).$$

The map $a \leftrightarrow P(\mu_{\alpha,a} e_j)(a)$ is a C^∞ map of Ω (and not just U) into R^s since

$$\mu_{\alpha,a} = \sum_{\beta \leqq \alpha} \binom{\alpha}{\beta} (-a)^{\alpha-\beta} \mu_{\beta,0},$$

and, by assumption, $P(\mu_{\beta,0} e_j)$ is a C^∞ map of Ω into R^s. We deduce:

3.3.7 If (3.3.5) holds, then there exist C^∞ maps a_α of Ω into the space of $s \times r$ matrices such that, for all $u \in C^\infty(U, r)$, we have

$$(Pu)(x) = \sum_{|\alpha| \leqq m} a_\alpha(x)(D^\alpha u)(x), \qquad x \in U.$$

We can now prove the theorem. It follows from proposition 3.3.2 that if $\Omega' \Subset \Omega$, there exist finitely many points $x_1, \ldots, x_N \in \overline{\Omega'}$ and constants $C, m > 0$ such that

$$\|Pu\|_0 \leqq C\|u\|_m$$

for all

$$u \in C_0^\infty(\Omega' - \bigcup \{x_\nu\}, r).$$

By (3.3.7), there exist C^∞ maps a_α of Ω' into R^s such that

$$(Pu)(x) = \sum_{|\alpha| \leqq m} a_\alpha(x)(D^\alpha u)(x), \qquad x \in \Omega' - \{x_1\} - \ldots - \{x_N\}.$$

Since both sides of this equation are continuous in Ω', we obtain equation 3.3.4.

The next theorem is a rewording of the theorem of Peetre.

3.3.8 THEOREM. Let V be a C^∞ manifold and let ξ, η be C^∞ vector bundles of rank r and s respectively. Let P be a linear differential operator from ξ to η. Then every point $a \in V$ has a neighbourhood U diffeomorphic is an open set in R^n such that ξ and η are trivial over U and the induced operator from ϑ_r to ϑ_s over Ω has the form

$$\sum_{|\alpha| \leq m} a_\alpha(x) D^\alpha,$$

with C^∞ $s \times r$ matrices a_α.

Let V be a C^∞ manifold, $a \in V$ and let $\boldsymbol{m}_a$ be the ring of germs of C^∞ functions vanishing at a (see definition following definition 2.1.8). Let ξ, η be C^∞ vector bundles on V and P a linear differential operator from ξ to η. We denote the fibres of ξ, η at a point $a \in V$ by ξ_a, η_a respectively.

3.3.9 Definition. The order $o_P(a)$ of P at $a \in V$ is the largest integer m such that $P(f^m s)(a) \neq 0$ for some $f \in \boldsymbol{m}_a$ and some section $s \in C^\infty(V, \xi)$. The order of P is simply the maximum $\max_{a \in V} o_P(a)$.

3.3.10 Remark. It is easy to verify that the order of an operator given by (3.3.4) is the largest integer m for which there exists an α with $|\alpha| = m$ and $a_\alpha \not\equiv 0$.

The next proposition is due to Hörmander [1964] and has proved to be of considerable importance for further developments because it leads directly to the theory of pseudo-differential operators.

Since, locally, any vector bundle is isomorphic to the trivial bundle, we may speak of the local uniform convergence, together with all partial derivatives, of a sequence of elements $s_\nu \in C^\infty(V, \xi)$. A C linear map $L: C^\infty(V, \xi) \to C^\infty(V, \eta)$ will be called weakly continuous if, for any sequence $\{s_\nu\}$ of sections in $C^\infty(V, \xi)$ which converges locally uniformly together with all partial derivatives to $s \in C^\infty(V, \xi)$, the sequence $\{Ls_\nu\}$ converges uniformly on compact subsets to Ls.

3.3.11 Theorem. A weakly continuous C-linear map

$$L: C^\infty(V, \xi) \to C^\infty(V, \eta)$$

is a linear differential operator of order $\leq m$ if and only if the following condition is fulfilled:

For any $s \in C^\infty(V, \xi)$, $a \in V$, and any real-valued C^∞ function f on V, the function

$$\kappa(\lambda)(a) = e^{-i\lambda f(a)}\{L(se^{i\lambda f})(a)\}$$

is a polynomial in λ of degree $\leqq m$ with values in η_a.

Proof. The fact that $\kappa(\lambda)$ is a polynomial of degree $\leqq m$ for any linear differential operator of order $\leqslant m$ is immediately checked by calculating in local coordinates.

To prove the converse, we proceed as follows. We write

$$\kappa(\lambda)(a) = \sum_{\nu=0}^{m} \kappa_\nu(f, s)(a)\lambda^\nu.$$

Then the map

$$a \mapsto \kappa_\nu(f, s)(a)$$

defines an element

$$\kappa_\nu(f, s) \in C^\infty(V, \eta).$$

If now $\lambda_1, \ldots, \lambda_k$ are real and $f_1, \ldots, f_k$ are real valued functions, we set

$$e^{-it(\lambda_1 f_1 + \ldots + \lambda_k f_k)} L(se^{it(\lambda_1 f_1 + \ldots + \lambda_k f_k)}) = \sum_{\nu=0}^{m} \kappa_\nu(\lambda, f, s)t^\nu,$$

where $\lambda = (\lambda_1, \ldots, \lambda_k)$, $f = (f_1, \ldots, f_k)$.

We see that

$$\kappa_\nu(\lambda, f, s) \in C^\infty(V, \eta)$$

and, for fixed λ and $u > 0$, we have

$$\kappa_\nu(u\lambda, f, s) = u^\nu \kappa_\nu(\lambda, f, s).$$

It follows (for example from Taylor's formula) that $\kappa_\nu(\lambda, f, s)(a)$ is a homogeneous polynomial of degree ν in $\lambda_1, \ldots, \lambda_k$.

Let now $a, b \in V$, $a \neq b$, and let U be a neighbourhood of a, b and $\varphi\colon U \to \Omega$ a C^∞ diffeomorphism of U onto an open set Ω in R^n (not necessarily connected).

For any $\beta \in C_0^\infty(\Omega)$, we have, by the inversion formula 3.2.2,

$$\beta(y) = (2\pi)^{-n/2} \int_{R^n} \hat{\beta}(\lambda) e^{i\langle y, \lambda \rangle} d\lambda,$$

where $\hat{\beta}$ is the Fourier transform of β. Hence, if $s \in C^\infty(V, \xi)$, and we set $s_0(x) = s(x)\beta(\varphi(x))$, we have

$$s_0(x) = (2\pi)^{-n/2} \int_{R^n} \hat{\beta}(\lambda) e^{i(\lambda_1\varphi_1(x)+\ldots+\lambda_n\varphi_n(x))} s(x) d\lambda_1 \ldots d\lambda_n.$$

Since L is weakly continuous, we deduce that

$$\begin{aligned} L(s_0)(a) &= (2\pi)^{-n/2} \int_{R^n} \hat{\beta}(\lambda) L\{se^{i(\lambda_1\varphi_1+\ldots+\lambda_n\varphi_n)}\}(a) d\lambda \\ &= (2\pi)^{-n/2} \sum_{\nu=0}^{m} \int_{R^n} \hat{\beta}(\lambda)\kappa_\nu(\lambda, \varphi, s)(a) e^{i\langle\lambda, \varphi(a)\rangle} d\lambda \\ &= \sum_{\nu=0}^{m} \sum_{|\alpha|=\nu} \sigma_{\alpha,\nu}(a)(2\pi)^{-n/2} \int_{R^n} \hat{\beta}(\lambda) e^{i\langle\lambda, \varphi(a)\rangle} \lambda^\alpha d\lambda, \end{aligned}$$

where

$$\sigma_{\alpha,\nu} \in C^\infty(V, \eta),$$

since, as has already been remarked, $\kappa_\nu(\lambda, \varphi, s)(a)$ is a homogeneous polynomial in λ of degree ν with values in η_a. By the inversion formula, this gives us

3.3.12 $$L(s_0)(a) = \sum_{\nu=0}^{m} \sum_{|\alpha|=\nu} \sigma_{\alpha,\nu}(a) i^{-|\alpha|} (D^\alpha\beta)(\varphi(a)).$$

We deduce immediately that if $s \in C_0^\infty(V, \xi)$ and has support in a small neighbourhood of a, then

$$\operatorname{supp}(Ls) \subset \operatorname{supp}(s);$$

in fact if $b \notin \operatorname{supp}(s)$, we have only to apply (3.3.12) with a function β so chosen that $\beta(\varphi(x)) = 0$ if x is near b, and $= 1$ if x is in a neighbourhood of supp (s). It follows that L is a linear differential operator. That L has order $\leq m$ follows also easily from (3.3.12).

Let V be a C^∞ manifold and ξ and η two C^∞ vector bundles on V. Let P be a linear differential operator from ξ to η, and suppose that P is of order m. Let $T^*(V)$ be the *real* cotangent bundle of V (§ 2.2) and $p^*: T^*(V) \to V$ the natural projection. Let $\xi = (E, p, V)$, $\eta = (F, q, V)$ and

$$E \times_V T^*(V) = \{(e, \omega) \in E \times T^*(V) \mid p(e) = p^*(\omega)\}.$$

[In other words, this consists of $a \in V$, a cotangent vector at a and an element of the fibre E_a over a.] We define a map

$$\sigma: E \times_V T^*(V) \to F$$

as follows. If $a \in V$ and $\omega \in T_a^*(V)$, let $f \in \boldsymbol{m}_a$ be a C^∞ function inducing ω, i.e. $\omega = (\mathrm{d}f)(a)$. Let $e \in E_a$, and let $s \in C_0^\infty(V, \xi)$ be so that $s(a) = e$ (such an s always exists). We define

$$\sigma(e, \omega) = P(f^m s)(a).$$

Note that, if $s(a) = 0$, then $f^m s$ has a 'zero of order $> m$' at a, hence $P(f^m s)(a) = 0$ in this case. Thus, $\sigma(e, \omega)$ depends only on e, not on the s with $s(a) = e$ chosen. Moreover, if $g \in \boldsymbol{m}_a$ and $\mathrm{d}(f-g)(a) = 0$, then $f \equiv g \bmod \boldsymbol{m}_a^2$, and we see that $f^m - g^m \in \boldsymbol{m}_a^{m+1}$. Hence, for any $s \in C_0^\infty(V, \xi)$,

$$P(f^m s)(a) = P(g^m s)(a).$$

3.3.13 DEFINITION. The map $\sigma = \sigma_P: E \times_V T^*(V) \to F$ so defined is called the *symbol* of the operator P.

3.3.14 DEFINITION. A linear differential operator P from ξ to η is called *elliptic* if for every $\omega \in T_a^*(V)$, $\omega \neq 0$, the $\mathbf{C}$ linear map

$$\sigma_a(\omega): E_a \to F_a, \qquad e \mapsto \sigma(e, \omega)$$

is injective. We shall also speak of P, in this case, simply as an elliptic operator from ξ to η.

3.3.15 EXAMPLES. (a) Let V be a C^∞ manifold of dimension n and let $\mathscr{A}^p = \mathscr{A}^p(V)$ be the set of C^∞ differential p-forms on V (with complex values). Moreover, if $\mathscr{E}^p = \wedge^p \mathfrak{T}^*(V)$ is the bundle of complex-valued forms on V, then $\mathscr{A}^p = C^\infty(V, \mathscr{E}^p)$ (theorem 2.2.8). Exterior differentiation

$$d: \mathscr{A}^p \to \mathscr{A}^{p+1}$$

is a differential operator of order 1 from $\mathscr{E}^p$ to $\mathscr{E}^{p+1}$. The symbol of the operator $d: \mathscr{A}^0 \to \mathscr{A}^1$ is given by

$$\sigma(e, \omega) = e\omega, \qquad e \in E_a^0, \qquad \omega \in T_a^*(V)$$

(note that $\mathscr{E}_a^0$ is canonically $\simeq \mathbf{C}$, so that the product above makes sense). We deduce at once that $d\colon \mathscr{A}^0 \to \mathscr{A}^1$ is elliptic.

(b) Let V be a complex manifold of complex dimension n and let $\mathscr{A}^{p,q}$ denote the space of C^∞ differential forms of type (p, q) on V. If $\mathscr{E}^{p,q}$ is the bundle of forms of type (p, q), then $\mathscr{A}^{p,q} = C^\infty(V, \mathscr{E}^{p,q})$ (§ 2.4). The operator $\bar\partial\colon \mathscr{A}^{p,q} \to \mathscr{A}^{p,q+1}$ is a differential operator of order 1 from $\mathscr{E}^{p,q}$ to $\mathscr{E}^{p,q+1}$. If $q = 0$, the symbol of $\bar\partial$ is defined by

$$\sigma(e, \omega) = \omega_0 \wedge e, \qquad e \in \mathscr{E}_a^{p,0}, \qquad \omega \in T_a^*(V),$$

where ω_0 is the projection of ω on $\mathscr{E}^{0,1}(V, a) \subset \mathfrak{T}_a^*(V)$. Note that if ω is a real covector, then the map $\omega \mapsto \omega_0$ is injective. Since e is of type $(p, 0)$ and ω_0 of type $(0, 1)$, $\omega_0 \wedge e = 0$ if and only if $\omega_0 = 0, e = 0$. It follows that $\bar\partial\colon \mathscr{A}^{p,0} \to \mathscr{A}^{p,1}$ *is elliptic.*

(c) It is trivial to check that the operator from ϑ_r to ϑ_r on an open set in R^n given by

$$(u_1, \ldots, u_r) \mapsto (\Delta u_1, \ldots, \Delta u_r),$$

$$\Delta u = \frac{\partial^2 u}{\partial x_1^2} + \ldots + \frac{\partial^2 u}{\partial x_n^2}$$

is elliptic. This is called the Laplace operator. We shall denote this again by Δ.

Let V be an *orientable* C^∞ manifold countable at infinity. Let $\mathfrak{T}^*(V)$ denote the bundle of complex covectors, and let

$$\mathscr{E}^n = \mathscr{E}^n(V) = \wedge^n \mathfrak{T}^*(V).$$

If $\xi = (E, p, V)$ is any vector bundle on V, and ξ^* is its dual bundle, we set

$$\xi' = \xi^* \otimes_{\mathbf{C}} \mathscr{E}^n$$

and call it the transpose of ξ. The natural pairing between a vector space and its dual defines, for each $a \in V$, a map

$$B_a\colon E_a' \times E_a \to \mathscr{E}_a^n, \qquad \xi' = (E', p', V).$$

This gives us a map

$$B\colon C^\infty(V, \xi') \times C^\infty(V, \xi) \to C^\infty(V, \mathscr{E}^n)$$

as follows. If $s' \in C^\infty(V, \xi')$, $s \in C^\infty(V, \xi)$, then

$$B(s', s)(a) = B_a(s'(a), s(a)).$$

If supp $(s) \cap$ supp (s') is compact, then $B(s', s)$ is a C^∞ n-form on V with compact support, and we define

$$\langle s', s\rangle_\xi = \langle s', s\rangle = \int_V B(s', s).$$

The scalar product $\langle s, s'\rangle_{\xi'}$ is defined (remark 3.3.16, (a)) and we have $\langle s, s'\rangle_{\xi'} = \langle s', s\rangle_\xi$.

3.3.16 REMARKS. (a) We have

$$(\xi')' = (\xi')^* \otimes \mathscr{E}^n = \xi \otimes (\mathscr{E}^n)^* \otimes \mathscr{E}^n,$$

which is canonically isomorphic to ξ since $\mathscr{E}^n$ is a line bundle and we can apply our remark at the end of § 3.1 (that $\eta \otimes \eta^*$ is canonically trivial for a line bundle η).

(b) If ξ is trivial, $\xi = (E, p, V)$ and

$$h: \xi \to V \times C^q$$

is an isomorphism, we have an isomorphism

$$h^*: \xi^* \to V \times C^q$$

defined as the inverse transpose of h (i.e. $h_a^* = {}^t h_a^{-1}$ for every $a \in V$). Moreover, if $x \in E_a$, $y^* \in E_a^*$, then

$$y^*(x) = \sum_{j=1}^{q} x_j y_j,$$

where $h(x) = a \times (x_1, \ldots, x_q)$, $h^*(y^*) = a \times (y_1, \ldots, y_q)$. We shall also identify $h(x)$ with its projection $(x_1, \ldots, x_q)$ on C^q.

3.3.17 PROPOSITION. Let P be a linear differential operator from ϑ_r to ϑ_s on an open set $\Omega \subset R^n$ given by

$$(Pu)(x) = \sum_{|\alpha| \leq m} a_\alpha(x) D^\alpha u(x).$$

Then there is a unique linear differential operator P^* from ϑ_s to ϑ_r, called the formal adjoint of P, such that

$$\int_\Omega (Pu(x), v(x))dx = \int_\Omega (u(x), P^*v(x))dx$$

for all $u \in C_0^\infty(\Omega, r)$, $v \in C_0^\infty(\Omega, s)$; here the scalar product (u_1, u_2) between vectors in C^r is defined by

$$(u_1, u_2) = \sum_{\nu=1}^{r} u_{1\nu}\bar{u}_{2\nu}, \qquad u_j = (u_{j1}, \ldots, u_{jr}).$$

Moreover, P^* is given by the formula

$$(P^*v)(x) = \sum_{|\alpha| \leq m} (-1)^{|\alpha|} D^\alpha(\overline{{}^t a_\alpha(x)} v(x));$$

here ${}^tA$ where A is a matrix denotes its transpose, and $\bar{A}$ the matrix whose elements are the complex conjugates of those of A. This follows easily from the formula

$$\int_\Omega \varphi(x)\overline{D^\alpha\psi(x)}dx = (-1)^{|\alpha|}\int_\Omega D^\alpha\varphi(x)\overline{\psi(x)}dx,$$

where $\varphi, \psi \in C_0^\infty(\Omega)$.

3.3.18 THEOREM. Let P be a linear differential operator from ξ to η, where ξ, η are vector bundles on the manifold V. Then there is a unique linear differential operator P' from η' to ξ' such that

3.3.19 $$\langle s, P't' \rangle_{\xi'} = \langle Ps, t' \rangle_{\eta'}$$

for all $s \in C_0^\infty(V, \xi)$, $t' \in C_0^\infty(V, \eta')$.

PROOF. It is obvious that if $t' \in C_0^\infty(V, \eta')$, then a section

$$P't' \in C_0^\infty(V, \xi'),$$

satisfying (3.3.19) for all $s \in C_0^\infty(V, \xi)$ is uniquely determined. Moreover, if t' vanishes on an open set $U \subset V$, then $\langle Ps, t' \rangle_{\eta'} = 0$ for all s with supp $(s) \subset U$, so that, if (3.3.19) holds, $\langle s, P't' \rangle_{\xi'} = 0$ for all s with supp $(s) \subset U$, and it follows that $P't'$ vanishes on U. Hence, it suffices to prove the existence of P' when V is an open set $\Omega \subset R^n$ and ξ and η are trivial on Ω.

Let

$$h_\xi : \xi \to \Omega \times C^r, \qquad h_\eta : \eta \to \Omega \times C^s$$

be isomorphisms and let h_ξ^* and h_η^* be the corresponding isomorphisms of the duals (remark 3.3.16, (b)). In terms of these isomorphisms, let P be given by

$$(Pu)(x) = \sum_{|\alpha| \leq m} a_\alpha(x) D^\alpha u(x).$$

If $\mathrm{d}x_1 \wedge \ldots \wedge \mathrm{d}x_n$ denotes the standard n-form on Ω, any element

$$t'_a \in F'_a = F_a^* \otimes \mathscr{E}_a^n$$

can be written uniquely in the form

$$t'_a = g_a \otimes (\mathrm{d}x_1 \wedge \ldots \wedge \mathrm{d}x_n)_a, \qquad g_a \in F_a^*.$$

If $t' \in C_0^\infty(\Omega, \eta')$, $t' = g \otimes (\mathrm{d}x_1 \wedge \ldots \wedge \mathrm{d}x_n)$, set

$$P't' = f \otimes (\mathrm{d}x_1 \wedge \ldots \wedge \mathrm{d}x_n).$$

Here f is defined by

$$h_\xi^*(f) = \overline{P^*}(h_\eta^*(g));$$

P^* is the formal adjoint of P as in proposition 3.3.17 and, for any operator

$$L = \sum c_\alpha(x) D^\alpha,$$

we have set

$$\bar{L} = \sum \overline{c_\alpha(x)} D^\alpha.$$

If $s \in C_0^\infty(\Omega, \xi)$ and t' is as above, we have

$$\begin{aligned} \langle Ps, t' \rangle_{\eta'} &= \int_\Omega (Ph_\xi(s), \overline{h_\eta^*(g)}) \mathrm{d}x_1 \wedge \ldots \wedge \mathrm{d}x_n \\ &= \int_\Omega (h_\xi(s), P^* \overline{h_\eta^*(g)}) \mathrm{d}x_1 \wedge \ldots \wedge \mathrm{d}x_n = \langle s, P't' \rangle_{\xi'}. \end{aligned}$$

3.3.20 Definition. The operator P' defined above is called the *(formal) transpose* of P.

3.3.21 Remark. If V is a real analytic manifold and ξ and η are real analytic bundles on V, an operator $P: C^\infty(V, \xi) \to C^\infty(V, \eta)$ is said to have analytic coefficients if Ps is analytic whenever s is an analytic section of ξ over an open set in V. It is easily seen that this is the case if and only if, in terms of local coordinates on V and (analytic) isomorphisms of ξ, η with trivial bundles, P is given by

$$(Pu)(x) = \sum_{|\alpha| \leqq m} a_\alpha(x) D^\alpha u(x),$$

where the a_α are real analytic maps into the space of $s \times r$ matrices. Moreover, the transpose P' then has again analytic coefficients.

3.3.22 REMARK. If P is an elliptic operator from ξ to η and if rank (ξ) = rank (η), then P' is again elliptic.

Note that the condition on the rank is necessary, since if there exists an elliptic operator from the bundle ξ to the bundle η, then rank $(\xi) \leqq$ rank (η).

Theorem 3.3.3 is due to PEETRE [1960], theorem 3.3.11 to HÖRMANDER [1965].

§ 3.4 The Sobolev spaces

Let Ω be an open set in R^n, p a real number $\geqq 1$ and q, m integers with $q \geqq 1$, $m \geqq 0$. Let

$$f = (f_1, \ldots, f_q) \colon \Omega \to C^q$$

be a C^∞ map. Consider the space of all those C^∞ maps $f\colon \Omega \to C^q$ for which

$$\sum_{|\alpha| \leqq m} \sum_{j=1}^{q} \int_\Omega |D^\alpha f_j(x)|^p \, dx < \infty.$$

We define a norm $|f|_{m,p}$ on this space by

$$|f|_{m,p}^p = \sum_{|\alpha| \leqq m} \sum_{j=1}^{q} \int_\Omega |D^\alpha f_j(x)|^p \, dx.$$

(Note that the triangle inequality follows from Minkowski's inequality.) We shall write $|f|_{m,p}^\Omega$ for this norm when its dependence on Ω is relevant.

3.4.1 DEFINITION. The completion of the above space relative to the norm $|f|_{m,p}$ is called the *Sobolev space* $H_{m,p}(\Omega)$. The completion of the space $C_0^\infty(\Omega, q)$ of C^∞ maps of Ω into C^q with compact support relative to $|f|_{m,p}$ is denoted by $\mathring{H}_{m,p}(\Omega)$.

Let $v_1, v_2 \in C^q$, $v_j = (v_{j1}, \ldots, v_{jq})$. We set, as usual,

$$(v_1, v_2) = \sum_{\nu=1}^{q} v_{1\nu}\overline{v_{2\nu}}.$$

If $f = (f_1, \ldots, f_q)\colon \Omega \to C^q$ and $g = (g_1, \ldots, g_q)\colon \Omega \to C^q$ are measurable mappings, we set

$$\langle f, g\rangle = \sum_{\nu=1}^{q} \int_\Omega f_\nu(x)\overline{g_\nu(x)}\mathrm{d}x = \int_\Omega (f(x), g(x))\mathrm{d}x$$

provided that the products $f_\nu(x)\overline{g_\nu(x)}$ are all summable.

3.4.2 Remarks. Let $f = (f_1, \ldots, f_q)$, $f_\nu \in L^p(\Omega)$. We write this simply $f \in L^p(\Omega, q)$ or $f \in L^p$. If there exist functions $h^\alpha \in L^\kappa(\Omega, q)$, $1 \leqq \kappa \leqq \infty$, $|\alpha| \leqq m$ so that, for all $g \in C_0^\infty(\Omega, q)$ we have

$$\int_\Omega (f(x), D^\alpha g(x))\mathrm{d}x = (-1)^{|\alpha|}\int_\Omega (h^\alpha(x), g(x))\mathrm{d}x,$$

we say that f has weak derivatives of order up to m in L^κ. The h^α are called the weak derivatives of f. Note that the h^α, if they exist, are uniquely determined (upto sets of measure zero).

If now $\{f_\nu\}$ is a sequence of elements in $C^\infty(\Omega, q)$ which converges in $H_{m,p}(\Omega)$, then $\{D^\alpha f_\nu\}$ converges in $L^p(\Omega)$ for $|\alpha| \leqq m$. Let its limit, in L^p, be f^α. Then, we have

$$\langle f_\nu, D^\alpha g\rangle = (-1)^{|\alpha|}\langle D^\alpha f_\nu, g\rangle \to (-1)^\alpha\langle f^\alpha, g\rangle$$

for any $g \in C_0^\infty(\Omega, q)$. It follows that the f^α are weak derivatives of

$$f = f^0 = \lim f_\nu.$$

In particular, if $\{f_\nu\}$, $\{g_\nu\}$ are two sequences defining the same element of $H_{m,p}(\Omega)$, then

$$\lim D^\alpha f_\nu = \lim D^\alpha g_\nu \qquad \text{for} \quad |\alpha| \leqq m.$$

Thus, if $f \in H_{m,p}(\Omega)$, we can define $D^\alpha f$ by

$$D^\alpha f = \lim D^\alpha f_\nu \qquad \text{if} \ \ f_\nu \to f \ \text{in} \ H_{m,p}(\Omega), \qquad f_\nu \in C^\infty(\Omega, q).$$

Let now $0 \leqq m' \leqq m$ and $f \in H_{m,p}(\Omega)$. Then, there is a sequence $\{f_\nu\}, f_\nu \in C^\infty(\Omega, q)$ which is a Cauchy sequence with respect to $|g|_{m,p}$, defining f. Now $|g|_{m',p} \leqq |g|_{m,p}$, so that $\{f_\nu\}$ is a Cauchy sequence relative to $|g|_{m',p}$, and so defines an element $f' \in H_{m',p}(\Omega)$, which is clearly independent of the sequence $\{f_\nu\}$ defining f. We set

$$f' = i(f) = i_{m,m'}(f).$$

3.4.3 Proposition. The linear map $i: H_{m,p}(\Omega) \to H_{m',p}(\Omega)$ is an injection.

Proof. If $i(f) = 0$, and $\{f_\nu\}$ is a sequence in $C^\infty(\Omega, q)$ defining f, then $f_\nu \to 0$ in $L^p(\Omega)$. If $g \in C_0^\infty(\Omega, q)$ we have, for $|\alpha| \leqq m$

$$0 = \lim \langle f_\nu, D^\alpha g\rangle = (-1)^{|\alpha|} \lim \langle D^\alpha f_\nu, g\rangle = (-1)^{|\alpha|}\langle f^\alpha, g\rangle,$$

so that $f^\alpha = 0$ for $|\alpha| \leqq m$, so that $f = 0$ in $H_{m,p}(\Omega)$.

Of course

3.4.4 $$i(\mathring{H}_{m,p}(\Omega)) \subset \mathring{H}_{m',p}(\Omega).$$

Further, if $\varphi \in C_0^\infty(\Omega) = C_0^\infty(\Omega, 1)$, then for any $f \in H_{m,p}(\Omega)$, we have

$$\varphi f \in \mathring{H}_{m,p}(\Omega)$$

and we have

$$D^\alpha(\varphi f) = \sum_{\beta \leqq \alpha} \binom{\alpha}{\beta} D^\beta \varphi D^{\alpha-\beta} f \qquad \text{for} \quad |\alpha| \leqq m.$$

Moreover, the obvious inclusion $C_0^\infty(\Omega) \subset C_0^\infty(R^n)$ gives rise to a norm preserving map

$$\mathring{H}_{m,p}(\Omega) \to \mathring{H}_{m,p}(R^n).$$

We shall therefore often identify $\mathring{H}_{m,p}(\Omega)$ with a subspace of $\mathring{H}_{m,p}(R^n)$.

3.4.5 Poincaré's inequality. Let Ω be a bounded open set in R^n and let $f \in C_0^\infty(\Omega)$. Then, if M is large enough (depending only on Ω), we have

$$f(x) = \int_{-M}^{x_1} \frac{\partial f}{\partial x_1}(t, x_2, \ldots, x_n)dt,$$

and, by Hölder's inequality,

$$\|f\|_{L^p} \leqq C(\Omega)\left\|\frac{\partial f}{\partial x_1}\right\|_{L^p}.$$

It follows easily that for any $f \in \mathring{H}_{m,p}(\Omega)$, we have

$$|f|_{m,p} \leqq C(\Omega, m) \sum_{|\alpha|=m} |D^\alpha f|_{0,p}.$$

3.4.6 REMARK. The case which is of most interest for our purposes is when $p = 2$. In this case we write

$$H_{m,2}(\Omega) = H_m(\Omega) \quad \text{and} \quad \mathring{H}_{m,2}(\Omega) = \mathring{H}_m(\Omega),$$
$$|f|_{m,2} = |f|_m.$$

The norm $|f|_m$ is of course induced by a scalar product

$$[f, g]_m = [f, g] = \sum_{|\alpha|\leqq m} \sum_{j=1}^{q} \int_\Omega D^\alpha f_j(x)\overline{D^\alpha g_j(x)}dx.$$

Also $H_m(\Omega)$ and $\mathring{H}_m(\Omega)$ are Hilbert spaces. These spaces are a little easier to handle than the general spaces $H_{m,p}(\Omega)$, the reason being provided by Plancherel's theorem. If $m = 0$, $H_0(\Omega) = L^2(\Omega)$ and we write $\langle f, g\rangle$ for $[f, g]_0$.

3.4.7 PROPOSITION. Let $f \in \mathring{H}_m(R^n)$ and let $\hat{f}$ denote the Fourier transform of f. There exist constants $c_1, c_2 > 0$ so that

$$c_1\int_{R^n}(1+|\xi|^2)^m|\hat{f}(\xi)|^2d\xi \leqq |f|_m^2 \leqq c_2\int_{R^n}(1+|\xi|^2)^m|\hat{f}(\xi)|^2d\xi.$$

PROOF. It is sufficient to prove this inequality for $C_0^\infty(R^n, q)$ since this latter space is dense in $\mathring{H}_m(R^n)$. One sees easily that there exist constants $c_1, c_2 > 0$ with

3.4.8 $$c_1(1+|\xi|^2)^m \leqq \sum_{|\alpha|\leqq m} |\xi^\alpha|^2 \leqq c_2(1+|\xi|^2)^m.$$

Now, if $f = (f_1, \ldots, f_q)$, we have

$$|f|_m^2 = \sum_{|\alpha|\leqq m} \sum_{j=1}^{q} \int |D^\alpha f_j(x)|^2 dx = \sum_{|\alpha|\leqq m} \sum_{j=1}^{q} \int |\widehat{D^\alpha f_j}(\xi)|^2 d\xi$$
(theorem 3.2.8)
$$= \sum_{|\alpha|\leqq m} \sum_{j=1}^{q} \int |\xi^\alpha|^2|\hat{f}_j(\xi)|^2 d\xi$$

and the proposition follows from (3.4.8).

3.4.9 THEOREM. We have

$$\mathring{H}_m(R^n) = H_m(R^n) = \left\{ f \in L^2(R^n, q) \,\middle|\, \int_{R^n} (1+|\xi|^2)^m |\hat{f}(\xi)|^2 \, d\xi < \infty \right\}.$$

PROOF. Let $\varphi \in C_0^\infty(R^n, 1)$ be so that $\varphi(x) = 1$ for $|x| \leqq 1, 0 \leqq \varphi(x) \leqq 1$. Let $f \in H_m(R^n)$ and

$$f_\nu(x) = \varphi(x/\nu) f(x).$$

Then f_ν has compact support. Moreover

$$D^\alpha f_\nu = \sum_{\beta \leqq \alpha} \binom{\alpha}{\beta} D^\beta \varphi_\nu D^{\alpha-\beta} f, \qquad \varphi_\nu(x) = \varphi(x/\nu).$$

Clearly, as $\nu \to \infty$, $D^\beta \varphi_\nu$ is bounded and $\to 0$ at every point if $|\beta| \geqq 1$, and $D^\beta \varphi_\nu \to 1$ if $\beta = 0$. Hence $D^\alpha f_\nu \to D^\alpha f$ in $L^2(R^n, q)$, so that $f_\nu \to f$ in $H_m(R^n)$. Since $f_\nu \in \mathring{H}_m(R^n)$, we see at once that $\mathring{H}_m(R^n) = H_m(R^n)$. Now, by proposition 3.4.7, we have

$$\mathring{H}_m(R^n) \subset \left\{ f \in L^2(R^n, q) \,\middle|\, \int_{R^n} (1+|\xi|^2)^m |\hat{f}(\xi)|^2 \, d\xi < \infty \right\}.$$

Conversely, if

$$\int_{R^n} (1+|\xi|^2)^m |\hat{f}(\xi)|^2 \, d\xi < \infty$$

for some $f \in L^2(R^n, q)$, then

$$(1+|\xi|^2)^{m/2} \hat{f}(\xi) \in L^2(R^n, q).$$

Hence there is a sequence $\{\hat{g}_\nu\}$ of functions in $\mathscr{S}$,

$$\hat{g}_\nu(\xi) \to (1+|\xi|^2)^{m/2} \hat{f}(\xi)$$

in $L^2(R^n, q)$. Now

$$\hat{g}_\nu / (1+|\xi|^2)^{m/2} \in \mathscr{S},$$

so that, by the inversion formula 3.2.2, there is a $h_\nu \in \mathscr{S}$ whose Fourier transform is $\hat{g}_\nu/(1+|\xi|^2)^{m/2}$. Then $h_\nu \in H_m(R^n)$. Since

$$\int (1+|\xi|^2)^m |\hat{h}_\nu(\xi) - \hat{h}_\mu(\xi)|^2 \, d\xi = \int |\hat{g}_\nu(\xi) - \hat{g}_\mu(\xi)| d\xi \to 0$$

as $\mu, \nu \to \infty$, it follows, from proposition 3.4.7, that h_ν converges in $H_m(R^n)$. Clearly $h_\nu \to f$ in $L^2(R^n, q)$, and our theorem follows.

3.4.10 DEFINITION. We have seen that there is an injection of $H_{m,p}(\Omega)$ into $H_{0,p}(\Omega) = L^p(\Omega, q)$. Given $f \in L^p(\Omega, q)$, we say that f is *strongly differentiable in L^p up to order m* if, for any $\Omega' \Subset \Omega$, $f|\Omega'$ lies in the image of $H_{m,p}(\Omega')$. If $p = 2$, we simply say that f is strongly differentiable.

3.4.11 THEOREM. Let φ be a C^∞ function with compact support in R^n, $\varphi \geqq 0$, supp $(\varphi) \subset \{x| \, |x| < 1\}$ and $\int_{R^n} \varphi(x)\mathrm{d}x = 1$.
Let $\varphi_\varepsilon(x) = \varepsilon^{-n}\varphi(x/\varepsilon)$. Let Ω be an open set in R^n and, if $x \in R^n$ and $f \in L^p(\Omega)$, set

$$(\varphi_\varepsilon * f)(x) = \int_\Omega \varphi_\varepsilon(x-y)f(y)\mathrm{d}y.$$

Then:
(a) For any $f \in L^p(\Omega)$, $\varphi_\varepsilon * f \to f$ in $L^p(\Omega)$ as $\varepsilon \to 0$.
(b) If $f \in \mathring{H}_{m,p}(\Omega)$, then, for $|\alpha| \leqq m$,

$$D^\alpha(\varphi_\varepsilon * f) = \varphi_\varepsilon * D^\alpha f.$$

PROOF. We extend f to R^n by setting $f(x) = 0$ if $x \notin \Omega$. We have

$$(\varphi_\varepsilon * f - f)(x) = \int \varphi_\varepsilon(x-y)(f(y)-f(x))\mathrm{d}y$$
$$= \int_{|y| \leqq \varepsilon} \varphi_\varepsilon(-y)(f(y+x)-f(x))\mathrm{d}y.$$

By Hölder's inequality, we have, if $p > 1$,

$$|(\varphi_\varepsilon * f - f)(x)|^p \leqq \left(\int_{|y| \leqq \varepsilon} |\varphi_\varepsilon(y)|^{p'}\right)^{p/p'} \int_{|y| \leqq \varepsilon} |f(y+x)-f(y)|^p \mathrm{d}y,$$

where p' is defined by $1/p + 1/p' = 1$. [If $p = 1$, the first factor on the right is to be replaced by sup $|\varphi_\varepsilon(y)| = \varepsilon^{-n}$ sup $|\varphi(y)|$]. Integrating over R^n with respect to x and taking pth roots, we get

$$\|\varphi_\varepsilon * f - f\|_{L^p} \leqq \varepsilon^{-n/p}\|\varphi\|_{L^{p'}} \left\{\int_{|y| \leqq \varepsilon} \mathrm{d}y \int |f(x+y)-f(x)|^p \mathrm{d}x\right\}^{1/p}$$

(where $\|\varphi\|_{L^{p'}} = \sup|\varphi(y)|$ if $p = 1$). This gives

$$\|\varphi_\varepsilon * f - f\|_{L^p} \leqq$$
$$\varepsilon^{-n/p}\left(\int_{|y|\leqq\varepsilon} dy\right)^{1/p} \|\varphi\|_{L^{p'}} \sup_{|y|\leqq\varepsilon}\left(\int |f(x+y)-f(x)|^p dx\right)^{1/p}.$$
$$= C\|\varphi\|_{L^{p'}} \sup_{|y|\leqq\varepsilon}\left(\int |f(x+y)-f(x)|^p dx\right)^{1/p}.$$

Now the last term $\to 0$ as $\varepsilon \to 0$; this is obvious if f is continuous with compact support, and follows for any $f \in L^p(\Omega)$ since continuous functions with compact support are dense. This proves (a).

To prove (b), let $f \in \mathring{H}_{m,p}(\Omega)$ and let $\{f_\nu\}$ be a sequence of elements in $C_0^\infty(\Omega, q)$ converging to f in $\mathring{H}_{m,p}(\Omega)$. Then

$$D^\alpha(\varphi_\varepsilon * f)(x) = \int_\Omega D^\alpha\varphi_\varepsilon(x-y)f(y)dy = \lim_{\nu\to\infty}\int_{R^n}(D^\alpha\varphi_\varepsilon)(x-y)f_\nu(y)dy$$
$$= \lim_{\nu\to\infty}\int_{R^n}\varphi_\varepsilon(x-y)D^\alpha f_\nu(y)dy = \int_\Omega \varphi_\varepsilon(x-y)D^\alpha f(y)dy$$
$$= (\varphi_\varepsilon * (D^\alpha f))(x).$$

3.4.12 THEOREM. If $f \in H_{m,p}(\Omega)$ and, for $|\alpha| \leqq m$, $D^\alpha f$ is strongly differentiable up to order m' in L^p, then f is strongly differentiable up to order $m+m'$ in L^p.

PROOF. By multiplying f by a suitable C^∞ function with compact support, we may suppose that f has compact support. If φ_ε is defined as in theorem 3.4.11, then $\varphi_\varepsilon * f$ is a C^∞ function. Further for $|\alpha| \leqq m$, we have, by theorem 3.4.11, (b):

$$D^\alpha(\varphi_\varepsilon * f) = \varphi_\varepsilon * (D^\alpha f).$$

Now, since $D^\alpha f \in \mathring{H}_{m',p}(\Omega)$, we have, whenever $|\alpha| \leqq m$, $|\beta| \leqq m'$,

$$D^{\alpha+\beta}(\varphi_\varepsilon * f) = D^\beta(\varphi_\varepsilon * D^\alpha f) = \varphi_\varepsilon * D^\beta(D^\alpha f) \qquad \text{(theorem 3.4.11, (b))}.$$

Now, when $\varepsilon \to 0$, this converges in $L^p(\Omega, q)$ (by theorem 3.4.11, (a)), and it follows that $f \in H_{m+m',p}(\Omega)$.

§ 3.5 The lemmata of Rellich and Sobolev

3.5.1 PROPOSITION. Let Ω be a bounded open set in R^n and let φ be a C^∞ function with compact support in R^n,

$$\varphi(x) \geqq 0, \qquad \int_{R^n} \varphi(x)\mathrm{d}x = 1.$$

Let

$$\varphi_\varepsilon(x) = \varepsilon^{-n}\varphi(x/\varepsilon).$$

Let $p \geqq 1$ and let p' be defined by $1/p'+1/p = 1$ if $p > 1$. Then, for any $f \in \mathring{H}_{m,p}(\Omega)$, where $m \geqq 1$, we have

$$|\varphi_\varepsilon * f - f|^{R^n}_{m-1,p} \leqq A\varepsilon||\varphi||_{L^{p'}}|f|_{m,p},$$

where A is a constant depending only on Ω, m and p and $||\varphi||_{L^{p'}}$ stands for sup $|\varphi(x)|$ if $p = 1$.

PROOF. Let f be a C^∞ function in R^n with supp $(f) \subset \Omega$. We have, for $x, y \in R^n$,

$$f(x+y)-f(x) = \sum_{j=1}^n y_j \int_0^1 \frac{\partial f}{\partial x_j}(x+ty)\mathrm{d}t.$$

Hence, if

$$g_y(x) = f(x+y)-f(x),$$

Hölder's inequality gives

$$|g_y(x)|^p \leqq n^{p-1}\sum_{j=1}^n |y_j|^p \int_0^1 \left|\frac{\partial f}{\partial x_j}(x+ty)\right|^p \mathrm{d}t,$$

so that

$$\begin{aligned} ||g_y||^p_{L^p} &\leqq n^{p-1}\sum_{j=1}^n |y_j|^p \int_0^1 \mathrm{d}t \int_{R^n} \left|\frac{\partial f}{\partial x_j}(x+ty)\right|^p \mathrm{d}x \\ &= n^{p-1}\sum_{j=1}^n |y_j|^p \int_{R^n} \left|\frac{\partial f}{\partial x_j}(x)\right|^p \mathrm{d}x \leqq n^{p-1}\sum_{j=1}^n |y_j|^p |f|^p_{1,p}. \end{aligned}$$

Hence: For $|y| \leqq \varepsilon$, we have

3.5.2 $$||g_y||_{L^p} \leqq n\varepsilon|f|_{1,p}.$$

Now

$$\varphi_\varepsilon * f(x) - f(x) = \int_{R^n} \varphi_\varepsilon(y)(f(x-y)-f(x))\mathrm{d}y.$$

Since

$$\operatorname{supp}(\varphi_\varepsilon) \subset \{x|\ |x| \leqq \varepsilon\},$$

this gives, if $p > 1$,

$$|\varphi_\varepsilon * f(x) - f(x)| \leqq \left(\int |\varphi_\varepsilon(y)|^{p'}\mathrm{d}y\right)^{1/p'} \left(\int_{|y| \leqq \varepsilon} |f(x+y)-f(x)|^p \mathrm{d}y\right)^{1/p}$$
$$= \varepsilon^{-n/p} ||\varphi||_{L^{p'}} \left(\int_{|y| \leqq \varepsilon} |g_y(x)|^p \mathrm{d}y\right)^{1/p},$$

and this inequality is clearly valid even if $p = 1$. Hence, by (3.5.2), we have

$$\left(\int_{R^n} |\varphi_\varepsilon * f(x) - f(x)|^p \mathrm{d}x\right)^{1/p} \leqq \varepsilon^{-n/p} ||\varphi||_{L^{p'}} n\varepsilon |f|_{1,p} \left(\int_{|y| \leqq \varepsilon} \mathrm{d}y\right)^{1/p}$$
$$= A\varepsilon ||\varphi||_{L^{p'}} |f|_{1,p}.$$

i.e.

$$|\varphi_\varepsilon * f - f|_{0,p}^{R^n} \leqq A\varepsilon ||\varphi||_{L^{p'}} |f|_{1,p}.$$

Applying this to the derivatives $D^\alpha f$ with $|\alpha| \leqq m-1$ and using the fact that $C_0^\infty(\Omega, q)$ is dense in $\mathring{H}_{m,p}(\Omega)$, we obtain the required inequality.

3.5.3 PROPOSITION. Let Ω be a bounded open set in R^n and k, a continuous function with compact support in R^n. Then, for $f \in L^p(\Omega)$, the function Kf defined by

$$(Kf)(x) = \int_\Omega k(x-y)f(y)\mathrm{d}y,$$

belongs to $L^p(R^n)$ and the operator

$$K: L^p(\Omega) \to L^p(R^n)$$

is completely continuous ($p \geqq 1$).

PROOF. The first statement is obvious since, for $f \in L^p(\Omega)$, Kf is continuous and has support in the compact set

$$S = \{x+y | x \in \overline{\Omega} \quad \text{and} \quad y \in \operatorname{supp}(k)\}.$$

Hence, given a sequence $\{f_\nu\}$ of elements in $L^p(\Omega)$ with $||f_\nu||_{L^p} \leqq 1$, it suffices to show that there is a subsequence $\{\nu_r\}$ so that $\{Kf_{\nu_r}\}$ converges uniformly on S. By Ascoli's theorem, it suffices, for this purpose, to show that the family $\{Kf \mid ||f||_{L^p} \leqq 1\}$ is bounded and equicontinuous.

For $\delta > 0$, set

$$\eta(\delta) = \sup_{|a-b| \leqq \delta} |k(a)-k(b)|.$$

Then, we have, by Hölder's inequality,

$$|Kf(x)| \leqq ||k||_{L^{p'}} ||f||_{L^p} \leqq ||k||_{L^{p'}}$$

and

$$|Kf(x)-Kf(y)| \leqq \eta(|x-y|) \int_\Omega |f(t)| \mathrm{d}t \leqq C\eta(|x-y|) ||f||_{L^p}$$

(since Ω is bounded). This proves the proposition.

3.5.4 RELLICH'S LEMMA. Let Ω be a bounded open set in R^n and let $0 \leqq m' < m$. Then the natural injection

$$i \colon \mathring{H}_{m,p}(\Omega) \to \mathring{H}_{m',p}(\Omega)$$

is completely continuous.

PROOF. For any continuous operator

$$T \colon \mathring{H}_{m,p}(\Omega) \to \mathring{H}_{m',p}(R^n),$$

set

$$||T|| = \sup_{f \neq 0} \left(\frac{|Tf|^{R^n}_{m',p}}{|f|^{\Omega}_{m,p}} \right),$$

and let j be the composite of i with the isometry $\mathring{H}_{m',p}(\Omega) \to \mathring{H}_{m',p}(R^n)$. Let T_ε be the operator $f \mapsto \varphi_\varepsilon * f$, where φ_ε is as in proposition 3.5.1. Then, by this proposition we have

$$||T_\varepsilon - j|| \to 0 \quad \text{as} \quad \varepsilon \to 0.$$

Further, each T_ε is completely continuous by proposition 3.5.3. Since the uniform limit of completely continuous operators is again completely continuous (an easily proved fact), the theorem is proved.

3.5.5 Proposition. When $p = 2$, the above theorem can be proved more simply using Plancherel's theorem.

Proof. Let $f \in \mathring{H}_m(\Omega)$, $|f|_m \leqq 1$. Consider, for complex $\xi = (\xi_1, \ldots, \xi_n) \in C^n$, the function

$$\hat{f}(\xi) = (2\pi)^{-n/2} \int_\Omega f(x) e^{-i(x_1\xi_1 + \ldots + x_n\xi_n)} dx.$$

By Schwarz's inequality, if ξ lies in a compact subset S of C^n,

$$|\hat{f}(\xi)| \leqq C(S)|f|_m$$

and so is uniformly bounded. Further, $\hat{f}$ is clearly holomorphic on C^n. Hence, by theorem 1.1.3′, the sequence $\{\hat{f}_\nu\}$ contains a subsequence $\{\hat{f}_{\nu_r}\}$ converging uniformly on compact subsets of C^n. We claim that the corresponding sequence $\{f_{\nu_r}\}$ then converges in $H_{m-1}(\Omega)$. Let $\varepsilon > 0$ be given, and choose $M > 0$ so that $1 + |\xi|^2 > 1/\varepsilon$ for $|\xi| \geqq M$. Then, by proposition 3.4.7,

$$\begin{aligned} |f_{\nu_r} - f_{\nu_s}|_{m-1}^2 &\leqq c_2 \int_{R^n} (1+|\xi|^2)^{m-1} |\hat{f}_{\nu_r} - \hat{f}_{\nu_s}|^2 d\xi \\ &\leqq c_2 \varepsilon \int_{|\xi|>M} (1+|\xi|^2)^m |\hat{f}_{\nu_r} - \hat{f}_{\nu_s}|^2 d\xi \\ &\quad + A(\varepsilon) \int_{|\xi| \leqq M} |\hat{f}_{\nu_r} - \hat{f}_{\nu_s}|^2 d\xi, \end{aligned}$$

where $A(\varepsilon)$ is a constant depending only on ε and m. Since $\{\hat{f}_{\nu_r}\}$ converges uniformly on compact subsets of R^n, the last integral tends to 0 as $r, s \to \infty$, which gives

$$\begin{aligned} \overline{\lim_{r,s\to\infty}} |f_{\nu_r} - f_{\nu_s}|_{m-1}^2 &\leqq c_2 \varepsilon \overline{\lim_{r,s\to\infty}} \int_{R^n} (1+|\xi|^2)^m |\hat{f}_{\nu_r} - \hat{f}_{\nu_s}|^2 d\xi \\ &\leqq c_3 \varepsilon. \end{aligned}$$

This proves the proposition.

3.5.6 Proposition. Let Ω be a bounded open set in R^n and m an integer $\geqq 0$. Then, for every $M > 0$, there exists a constant $A > 0$ depending only on M, Ω and m such that, for all $f \in \mathring{H}_m(\Omega)$, we have

$$\int_{R^n} (1+|\xi|^2)^m |\hat{f}(\xi)|^2 d\xi \leqq A \int_{|\xi| \geqq M} (1+|\xi|^2)^m |\hat{f}(\xi)|^2 d\xi.$$

Proof. If the result is false, there exists a sequence $\{f_\nu\}_{\nu \geqq 1}$ of C^∞ functions with compact support in Ω such that

$$\int_{R^n} (1+|\xi|^2)^m |\hat{f}_\nu(\xi)|^2 \mathrm{d}\xi = 1,$$

$$\int_{|\xi| \geqq M} (1+|\xi|^2)^m |\hat{f}_\nu(\xi)|^2 \mathrm{d}\xi \to 0 \quad \text{as} \quad \nu \to \infty.$$

Again, if we let $\xi = (\xi_1, \ldots, \xi_n)$ be complex, the holomorphic functions

$$\hat{f}_\nu(\xi) = (2\pi)^{-n/2} \int_\Omega f_\nu(x) e^{-\mathrm{i}(x_1\xi_1 + \ldots + x_n\xi_n)} \mathrm{d}x$$

on C^n are uniformly bounded on compact sets, so that we may assume (theorem 1.1.3′) that $\hat{f}_\nu$ converges uniformly on compact subsets of C^n (in particular of R^n) to a holomorphic function g. Now

$$\int_{M \leqq |\xi| \leqq M+1} (1+|\xi|^2)^m |g(\xi)|^2 \mathrm{d}\xi$$
$$= \lim_{\nu \to \infty} \int_{M \leqq |\xi| \leqq M+1} (1+|\xi|^2)^m |\hat{f}_\nu(\xi)|^2 \mathrm{d}\xi = 0$$

by our assumption above. Hence $g(\xi) = 0$ for real ξ with $M \leqq |\xi| \leqq M+1$. Since g is holomorphic in C^n, this implies that $g \equiv 0$, so that $\hat{f}_\nu$ converges uniformly to zero on compact subsets of C^n. In particular,

$$\int_{|\xi| \leqq M} (1+|\xi|^2)^m |\hat{f}_\nu(\xi)|^2 \mathrm{d}\xi \to 0 \quad \text{as} \quad \nu \to \infty.$$

It follows, since

$$\int_{|\xi| \geqq M} (1+|\xi|^2)^m |\hat{f}_\nu(\xi)|^2 \mathrm{d}\xi \to 0$$

by hypothesis that

$$\int_{R^n} (1+|\xi|^2)^m |\hat{f}_\nu(\xi)|^2 \mathrm{d}\xi \to 0.$$

This is a contradiction.

3.5.7 Remarks. Poincaré's inequality 3.4.5 for $p = 2$ can be written, using Plancherel's theorem, as follows: There is a constant $C(\Omega, m)$ such that, for all $f \in \mathring{H}_m(\Omega)$, we have

$$\int_{R^n} (1+|\xi|^2)^m |\hat{f}(\xi)|^2 \mathrm{d}\xi \leqq C(\Omega, m) \int_{R^n} |\xi|^{2m} |\hat{f}(\xi)|^2 \mathrm{d}\xi.$$

Thus, proposition 3.5.6 can be looked upon as a sharper form of Poincaré's inequality.

It is further possible to obtain the best possible constants in the inequality 3.5.6. This is connected with rather interesting questions in Fourier analysis; see Fuchs [1964].

3.5.8 Proposition. (Polar coordinates in R^n). Let $R^+ = \{t \in R | t > 0\}$ and S^{n-1} be the $(n-1)$ sphere in R^n (example 2.5.6).
Let

$$\vartheta: R^+ \times S^{n-1} \to R^n - \{0\}$$

be the map $(t, x) \mapsto tx$. There exists an $(n-1)$-form ω on S^{n-1} such that

$$\vartheta^*(\mathrm{d}y_1 \wedge \ldots \wedge \mathrm{d}y_n) = t^{n-1} \mathrm{d}t \wedge \omega.$$

Moreover,

$$\int_{S^{n-1}} \omega \neq 0.$$

Proof. If $x_1, \ldots, x_n$ are the restrictions to S^{n-1} of the coordinate functions in R^n, we may take

$$\omega = \sum_{k=1}^{n} x_k \mathrm{d}x_1 \wedge \ldots \wedge \widehat{\mathrm{d}x_k} \wedge \ldots \wedge \mathrm{d}x_n$$

where the $\wedge$ over a term $\mathrm{d}x_k$ means that that term is to be omitted. The fact that

$$\int_{S^{n-1}} \omega \neq 0$$

follows from the fact that

$$\int_U \mathrm{d}y_1 \wedge \ldots \wedge \mathrm{d}y_n > 0$$

where

$$U = \vartheta(I \times S^{n-1}), \qquad I = \{t | \tfrac{1}{2} < t < 1\}.$$

3.5.9 SOBOLEV'S LEMMA. Let Ω be an open set in R^n and let $m > n/p$. Then for every compact set $K \subset \Omega$, there is a constant $C > 0$ such that for every $f \in C^\infty(\Omega)$, with supp $(f) \subset K$, we have

$$\sup_{x \in K} |f(x)| \leqq C|f|_{m,p}.$$

PROOF. We may suppose that $\Omega = R^n$. Further, given K, we can choose a compact set $L \subset R^n$ such that for any $y \in K$, the function

$$g = g_y : x \mapsto f(x+y)$$

has support $\subset L$. Thus, it is sufficient to show that for any $f \in C^\infty$ with supp $(f) \subset L$, we have

$$|f(0)| \leqq C|f|_{m,p}.$$

Let $\vartheta: R^+ \times S^{n-1} \to R^n - \{0\}$ be the map defined in proposition 3.5.8.

If $f \in C^\infty(R^n)$, let $g = f \circ \vartheta$. Then the partial derivatives $\partial^m g(t, x)/\partial t^m$ can be obtained in the following way. There are homogeneous polynomials q_α of degree m on R^n such that

3.5.10 $$\frac{\partial^m g(t, x)}{\partial t^m} = \sum_{|\alpha|=m} q_\alpha \left(\frac{y}{||y||}\right) D^\alpha f(y), \qquad y = \vartheta(t, x) = tx;$$

in particular, the functions $q_\alpha(y/||y||)$ are bounded. We have, if M is a large enough constant and $x \in S^{n-1}$,

$$f(0) = -\int_0^M \frac{\partial}{\partial t} f(tx)\mathrm{d}t = \frac{(-1)^m}{(m-1)!}\int_0^M t^{m-1}\frac{\partial^m g(t, x)}{\partial t^m}\mathrm{d}t.$$

Multiplying by ω and integrating over S^{n-1}, we obtain

$$f(0)\int_{S^{n-1}} \omega = C_m \int_{S^{n-1}}\int_0^M \frac{\partial^m g(t, x)}{\partial t^m} t^{m-1}\mathrm{d}t \wedge \omega$$

which gives, since

$$\int_{S^{n-1}} \omega \neq 0,$$

3.5.11 $$f(0) = C'_m \int_0^M \int_{S^{n-1}} t^{m-n}\frac{\partial^m g(t, x)}{\partial t^m} t^{n-1}\mathrm{d}t \wedge \omega$$

$$= C'_m \int_{||y|| \leqq M} t^{m-n} g_m(y)\mathrm{d}y,\ t = ||y||,$$

(by proposition 3.5.8) where $g_m(y)$ is given by (3.5.10). Hölder's inequality gives, if $p > 1$,

$$|f(0)| \leqq C' \left(\int_{\|y\| \leqq M} t^{(m-n)p'} \mathrm{d}y \right)^{1/p'} \left(\int_{\|y\| \leqq M} |g_m(y)|^p \mathrm{d}y \right)^{1/p},$$

where $1/p + 1/p' = 1$. Since $m > n/p$, we have $(m-n)p' + n - 1 > -1$, so that

$$\int_{\|y\| \leqq M} t^{(m-n)p'} \mathrm{d}y = \int_0^M t^{(m-n)p'+n-1} \mathrm{d}t \int_{S^{n-1}} \omega = C'' < \infty$$

Also, by (3.5.10),

$$\int_{\|y\| \leqq M} |g_m(y)|^p \mathrm{d}y \leqq \text{const } |f|_{m,p}^p,$$

and we obtain

$$|f(0)| \leqq C |f|_{m,p}$$

as required.

If $p = 1$, $m \geqq n$, the required inequality follows at once from (3.5.11).

3.5.12 COROLLARY. If Ω is an open set in R^n and K is a compact subset of Ω, then, for any $f \in C^\infty(\Omega, q)$, we have, for $m > n/p$,

$$\sup_{x \in K} |f(x)| \leqq C |f|_{m,p}.$$

We have only to apply lemma 3.5.9 to the components of φf, where $\varphi \in C_0^\infty(\Omega)$ and $\varphi(x) = 1$ for $x \in K$.

3.5.13 PROPOSITION. Let Ω be an open set in R^n and $m > n/p$. Then any $f \in H_{m,p}(\Omega)$ is equal, almost everywhere, to a function with continuous derivatives of order $\leqq m - [n/p] - 1$; here $[X]$ is the largest integer $\leqq X$.

PROOF. By multiplying f by a suitable function with compact support, we may suppose that $f \in \mathring{H}_{m,p}(\Omega)$ and that Ω is bounded. Let

$$f_\nu \in C_0^\infty(\Omega, q), \qquad |f_\nu - f|_{m,p} \to 0.$$

By corollary 3.5.12, if K is compact in Ω, there is $C > 0$ so that for $|\alpha| < m-(n/p)$,

$$\sup_{x\in K} |D^\alpha(f_\nu - f_\mu)(x)| \leqq C|D^\alpha(f_\nu - f_\mu)|_{m-|\alpha|,p}$$
$$\leqq C|f_\nu - f_\mu|_{m,p} \to 0 \quad \text{as} \quad \nu, \mu \to \infty.$$

Hence, for $|\alpha| < m-(n/p)$, $D^\alpha f_\nu$ converges uniformly on compact subsets of Ω. The proposition follows.

3.5.14 PROPOSITION. When $p = 1$ or $p = 2$, lemma 3.5.9 can be proved more simply.

Case $p = 1$. We have, if M is large,

$$f(x) = \int_{-M}^{x_1} \dots \int_{-M}^{x_n} \frac{\partial^n f(t_1, \dots, t_n)}{\partial t_1 \dots \partial t_n} \mathrm{d}t_1 \dots \mathrm{d}t_n,$$

so that

$$|f(x)| \leqq |f|_{n,1} \leqq |f|_{m,1} \qquad \text{if} \quad m \geqq n.$$

In fact, Hölder's inequality gives

3.5.15
$$|f(x)| \leqq C|f|_{m,p}$$

if $m \geqq n$ and $p \geqq 1$.

This inequality is usually sufficient.

Case $p = 2$. We have

$$f(x) = (2\pi)^{-n/2} \int_{R^n} \hat{f}(\xi) e^{i\langle x, \xi\rangle} \mathrm{d}\xi$$
$$= (2\pi)^{-n/2} \int_{R^n} e^{i\langle x, \xi\rangle}(1+|\xi|^2)^{-m/2}\hat{f}(\xi)(1+|\xi|^2)^{+m/2} \mathrm{d}\xi;$$

Schwarz' inequality gives

$$|f(x)|^2 \leqq C \int_{R^n} (1+|\xi|^2)^{-m} \mathrm{d}\xi \int_{R^n} |\hat{f}(\xi)|^2 (1+|\xi|^2)^m \mathrm{d}\xi$$
$$\leqq C'|f|_m^2$$

(by proposition 3.4.7) since for $m > n/2$

$$\int_{R^n} (1+|\xi|^2)^{-m} \mathrm{d}\xi < \infty.$$

A useful remark concerning the norms in $\mathring{H}_m(\Omega)$ is:

3.5.16 PROPOSITION. For any $\varepsilon > 0$, there is a constant $C(\varepsilon) > 0$ such that

$$|f|_{m-1}^2 \leqq \varepsilon |f|_m^2 + C(\varepsilon)|f|_0^2 \qquad \text{for all} \quad f \in \mathring{H}_m(\Omega).$$

PROOF. It suffices to prove this for all $f \in C_0^\infty(R^n)$. Now, by proposition 3.4.7,

$$|f|_{m-1}^2 \leqq c_2 \int_{R^n} (1+|\xi|^2)^{m-1} |\hat{f}(\xi)|^2 \,\mathrm{d}\xi.$$

Given $\varepsilon > 0$, there is a constant $C(\varepsilon) > 0$ so that

$$(1+|\xi|^2)^{m-1} \leqq \varepsilon c_2^{-1}(1+|\xi|^2)^m + C(\varepsilon)$$

for all $\xi \in R^n$. Hence

$$|f|_{m-1}^2 \leqq \varepsilon \int_{R^n} (1+|\xi|^2)^m |\hat{f}(\xi)|^2 \,\mathrm{d}\xi + C(\varepsilon) \int_{R^n} |\hat{f}(\xi)|^2 \,\mathrm{d}\xi,$$

and the proof is completed by proposition 3.4.7.

3.5.17 REMARK. This is equivalent to the following: For any $\varepsilon > 0$, there is $C(\varepsilon) > 0$ such that

$$|f|_{m-1} \leqq \varepsilon |f|_m + C(\varepsilon)|f|_0$$

for all $f \in \mathring{H}_m(\Omega)$.

Rellich's lemma remains true if we replace $\mathring{H}_{m,p}(\Omega)$ by $H_{m,p}(\Omega)$ if the boundary of Ω is sufficiently smooth (see RELLICH [1930]).

Several proofs of Sobolev's lemma are available. SOBOLEV [1938] obtained several very precise inequalities. However, most of these proofs are more complicated than the one we have given here.

§ 3.6 The inequalities of Gårding and Friedrichs

We shall consider, in this section, differential operators from ϑ_r to ϑ_s on an open set $\Omega \subset R^n$. We suppose the operator given in the form

$$(Pu)(x) = \sum_{|\alpha| \leqq m} a_\alpha(x) D^\alpha u(x), \qquad u \in C^\infty(\Omega, r).$$

We see immediately that if $v = (v_1, \ldots, v_r) \in C^r$, $x_0 \in \Omega$ and $f \in \mathfrak{m}_{x_0}$, then

$$P(f^m v)(x_0) = m! \sum_{|\alpha|=m} \xi_1^{\alpha_1} \ldots \xi_n^{\alpha_n} a_\alpha(x_0) v,$$

where

$$\xi_j = \frac{\partial f}{\partial x_j}(a).$$

If we set

$$p_P(x, \xi) = p(x, \xi) = p_m(x, \xi) = \sum_{|\alpha|=m} \xi^\alpha a_\alpha(x), \qquad x \in \Omega, \xi \in R^n,$$

then from the above remark and the definition 3.3.14, it follows that *P is elliptic if and only if, for any $\xi \neq 0$, $\xi \in R^n$, and $x \in \Omega$, the map*

$$p(x, \xi) \colon C^r \to C^s$$

is injective. This function p is called the characteristic polynomial of the operator P.

In case $r = s$, it is useful to consider a more special class of operators.

3.6.1 DEFINITION. A linear differential operator of order m from ϑ_r to itself on Ω is called *(uniformly) strongly elliptic* if there exists a constant $C > 0$ such that for all $\xi \in R^n$, $x \in \Omega$ and $v \in C^r$, we have

$$\operatorname{Re}(p(x, \xi)v, v) \geqq C|\xi|^m |v|^2.$$

If $n > 1$, then any strongly elliptic operator is of even order. In fact, if $x \in \Omega$ and $v \neq 0$, the function

$$Q(\xi) = \operatorname{Re}(p(x, \xi)v, v)$$

is a homogeneous polynomial of degree $m(=$ order of $P)$. It is clear that for almost all values of a, $b \in R^n$, the polynomial $Q(a+\lambda b)$ of the real variable λ has degree m in λ, so has a real zero if m is odd. If $n > 1$, we can choose a, b so that $a+\lambda b \neq 0$ for all $\lambda \in R$ and Q would have a real non-trivial zero.

Let P_1 be a linear differential operator of order m_1 from ϑ_r to ϑ_s, P_2 an operator of order m_2 from ϑ_s to ϑ_t. Then

$$P_2 \circ P_1 \colon C^\infty(\Omega, r) \to C^\infty(\Omega, t)$$

is a linear differential operator. It is easily checked that this operator has order $\leq m_1+m_2$.

If

$$p_1(x, \xi): C^r \to C^s, \; p_2(x, \xi): C^s \to C^t$$

denote the characteristic polynomials of P_1, P_2 respectively, then if $P_2 \circ P_1$ has order m_1+m_2, its characteristic polynomial is $p_2 \circ p_1$. This is the case if and only if $p_2 \circ p_1 \neq 0$. In particular, if P_1, P_2 are elliptic, then $p_2 \circ p_1: C^r \to C^t$ is clearly injective for $\xi \neq 0$, and it follows that $P_2 \circ P_1$ is again elliptic. Further, if P^* denotes the formal adjoint of P (proposition 3.3.17), its characteristic polynomial is given by

$$p^*(x, \xi) = (-1)^m \, {}^t\overline{p(x, \xi)},$$

(p = characteristic polynomial of P).

From these remarks we deduce immediately the following:

3.6.2 COROLLARY. If P is an elliptic operator from ϑ_r to ϑ_s on Ω, then, for any $\Omega' \Subset \Omega$, the operator $(-1)^m P^* \circ P$ is uniformly strongly elliptic on Ω' of even order from ϑ_r to itself.

PROOF. If

$$L = (-1)^m P^* \circ P$$

and p_L is its characteristic polynomial, then

$$p_L = (-1)^m p^* \circ p,$$

so that, for $v \in C^r$,

$$(p_L(x, \xi)v, v) = (p(x, \xi)v, p(x, \xi)v) = ||p(x, \xi)v||^2.$$

Now, if $x \in \Omega' \Subset \Omega$ and $|\xi| = 1$, $|v| = 1$, then $||p(x, \xi)v||^2$ is bounded below since P being elliptic, $p(x, \xi)v \neq 0$. It follows, by homogeneity, that

$$(p_L(x, \xi)v, v) \geqq C|\xi|^{2m}|v|^2.$$

3.6.3 GÅRDING'S INEQUALITY. Let P be a uniformly elliptic operator of (even) order $2m$ on Ω from ϑ_r to itself. Then, for any relatively compact open subset Ω' of Ω, there exist constants $C > 0$ and $B > 0$ such that

$$(-1)^m \operatorname{Re} \langle Pu, u \rangle \geqq C|u|_m^2 - B|u|_0^2$$

for all $u \in C_0^\infty(\Omega', r)$.

Proof. The theorem is proved in three steps.

Step I. Let P be given by

$$Pu(x) = \sum_{|\alpha| \leqq 2m} a_\alpha D^\alpha u(x),$$

where the a_α are constant matrices, (i.e. P is an operator with constant coefficients). Plancherel's theorem gives

$$\langle Pu, u \rangle = \langle \widehat{Pu}, \hat{u} \rangle.$$

Now

$$\widehat{Pu} = \sum_{|\alpha| \leqq 2m} a_\alpha \widehat{D^\alpha u}(\xi) = (-1)^m \sum_{|\alpha| = 2m} a_\alpha \xi^\alpha \hat{u}(\xi) + \sum_{|\alpha| \leqq 2m-1} a_\alpha i^{|\alpha|} \xi^\alpha \hat{u}(\xi)$$
$$= (-1)^m p(\xi)\hat{u}(\xi) + q(\xi)\hat{u}(\xi),$$

where p is the characteristic polynomial of P and q is a polynomial of degree $\leqq 2m-1$ (with coefficients $r \times r$ complex matrices). Thus

$$(-1)^m \operatorname{Re} \langle Pu, u \rangle$$
$$= \operatorname{Re} \int_{R^n} (p(\xi)\hat{u}(\xi), \hat{u}(\xi)) \mathrm{d}\xi + (-1)^m \operatorname{Re} \int_{R^n} (q(\xi)\hat{u}(\xi), \hat{u}(\xi)) \mathrm{d}\xi$$
$$\geqq c_1 \int_{R^n} |\xi|^{2m} |\hat{u}(\xi)|^2 \mathrm{d}\xi - c_2 \int_{R^n} (1+|\xi|)^{2m-1} |\hat{u}(\xi)|^2 \mathrm{d}\xi,$$

where $c_1, c_2 > 0$; we have used the fact that, P being uniformly strongly elliptic, there is a $c_1 > 0$ with

$$\operatorname{Re}(p(\xi)v, v) \geqq c_1 |\xi|^{2m} |v|^2, \qquad v \in C^r$$

and the fact that q is of degree $\leqq 2m-1$. Let M be so large that

$$c_1 |\xi|^{2m} - c_2 (1+|\xi|)^{2m-1} \geqq \tfrac{1}{2} c_1 (1+|\xi|^2)^m$$

for $|\xi| \geqq M$.

Then (if we set $c_3 = \frac{1}{2} c_1$),

$$(-1)^m \operatorname{Re} \langle Pu, u \rangle$$
$$\geqq c_3 \int_{|\xi| \geqq M} (1+|\xi|^2)^m |\hat{u}(\xi)|^2 \mathrm{d}\xi - c_2 \int_{|\xi| \leqq M} (1+|\xi|)^{2m-1} |\hat{u}(\xi)|^2 \mathrm{d}\xi$$
$$\geqq c_3 \int_{R^n} (1+|\xi|^2)^m |\hat{u}(\xi)|^2 \mathrm{d}\xi - c_4 \int_{R^n} |\hat{u}(\xi)|^2 \mathrm{d}\xi,$$

where

$$c_4 = \sup_{|\xi| \leqq M} \{c_3(1+|\xi|^2)^m + c_2(1+|\xi|)^{2m-1}\}.$$

By proposition 3.4.7, there is a constant $c > 0$ such that

$$(-1)^m \operatorname{Re} \langle Pu, u\rangle \geqq c|u|_m^2 - c_4|u|_0^2.$$

Step II. This consists of the following.

3.6.4 PROPOSITION. For any $x_0 \in \Omega$, there is a neighbourhood U of x_0 such that

$$(-1)^m \operatorname{Re} \langle Pu, u\rangle \geqq c|u|_m^2 - B|u|_0^2$$

for all $u \in C_0^\infty(U, r)$; here c, B are constants > 0 depending only on x_0.

PROOF. We can find linear differential operators $Q_1, \ldots, Q_N$; $R_1, \ldots, R_N$ of orders $\leqq m$ from ϑ_r to itself on Ω such that

$$P = \sum_{\nu=1}^{N} R_\nu^* \circ Q_\nu,$$

where R_ν^* denotes the formal adjoint of R_ν. Now, we can write

$$Q_\nu = Q_\nu^0 + Q_\nu', \qquad R_\nu = R_\nu^0 + R_\nu',$$

where Q_ν^0, R_ν^0 are operators with constant coefficients of order $\leqq m$ and *all coefficients of* Q_ν', R_ν' *vanish at* $x_0 \in \Omega$. We then have

$$\langle Pu, u\rangle = \langle P^0u, u\rangle + \sum_\nu \{\langle Q_\nu' u, R_\nu^0 u\rangle + \langle Q_\nu^0 u, R_\nu' u\rangle + \langle Q_\nu' u, R_\nu' u\rangle\},$$

where

$$P^0 = \sum_{\nu=1}^{N} (R_\nu^0)^* \circ Q_\nu^0.$$

Since the coefficients of Q_ν', R_ν' vanish at $x_0 \in \Omega$ and are of order $\leqq m$, we see at once that for any $\varepsilon > 0$, there is a neighbourhood U of x_0 such that the last sum above is, in absolute value $\leqq \varepsilon|u|_m^2$ for all $u \in C_0^\infty(U, r)$.

Applying the result of Step I to P^0, we obtain

$$(-1)^m \operatorname{Re} \langle Pu, u\rangle \geqq (c_0 - \varepsilon)|u|_m^2 - B|u|_0^2, \qquad u \in C_0^\infty(U, r)$$

if

$$(-1)^m \operatorname{Re} \langle P^0u, u\rangle \geqq c_0|u|_m^2 - B|u|_0^2.$$

Step III. This is the general case.

If $\Omega' \Subset \Omega$, we can find, by proposition 3.6.4 above, constants $c > 0$, $B > 0$ and a finite covering $U_1, \ldots, U_h$ of $\bar{\Omega}'$ such that

3.6.5 $$(-1)^m \operatorname{Re} \langle Pu, u\rangle \geqq c|u|_m^2 - B|u|_0^2 \quad \text{for} \quad u \in C_0^\infty(U_j, r), \quad j = 1, \ldots, h.$$

Let $\eta_j \in C_0^\infty(U_j, 1)$, $0 \leqq \eta_j \leqq 1$, and $\sum \eta_j^2(x) = 1$ for all x in a neighbourhood of $\bar{\Omega}'$ (the η_j exist by lemma 1.2.7).

We remark that for a suitable $C > 0$ depending only on the η_j's, we have

$$\left|\, |\eta_j u|_m^2 - \sum_{|\alpha| \leqq m} |\eta_j D^\alpha u|_0^2 \right| \leqq C|u|_{m-1}|u|_m$$

and

$$\langle P(\eta_j u), \eta_j u\rangle - \langle \eta_j Pu, \eta_j u\rangle = \langle Lu, u\rangle,$$

where L is a differential operator of order $\leqq 2m-1$. Hence (writing L in the form $\sum B_\mu^* \circ A_\mu$, where the A_μ's are operators of order $\leqq m$, B_μ's, operators of order $\leqq m-1$) we see that there is $C > 0$ so that

$$|\langle P(\eta_j u), \eta_j u\rangle - \langle \eta_j Pu, \eta_j u\rangle| \leqq C|u|_m|u|_{m-1}.$$

This gives us, for any $u \in C_0^\infty(\Omega', r)$:

$$\begin{aligned}
c|u|_m^2 &= c \sum_{|\alpha| \leqq m} |D^\alpha u|_0^2 = c \sum_{j=1}^h \sum_{|\alpha| \leqq m} |\eta_j D^\alpha u|_0^2 \\
&\leqq c \sum_{j=1}^h |\eta_j u|_m^2 + cC|u|_m|u|_{m-1} \\
&\leqq \sum_{j=1}^h (-1)^m \operatorname{Re} \langle P(\eta_j u), \eta_j u\rangle + B'|u|_0^2 + C'|u|_m|u|_{m-1} \\
&\leqq \sum_{j=1}^h (-1)^m \operatorname{Re} \langle \eta_j Pu, \eta_j u\rangle + B'|u|_0^2 + C''|u|_m|u|_{m-1} \\
&= (-1)^m \operatorname{Re} \langle Pu, u\rangle + B'|u|_0^2 + C''|u|_m|u|_{m-1}.
\end{aligned}$$

Now, if $\delta > 0$, we have

$$2|w_1 w_2| \leqq \delta|w_1|^2 + \frac{1}{\delta}|w_2|^2$$

for any complex numbers w_1, w_2. Hence

$$2C''|u|_m|u|_{m-1} \leqq \delta|u|_m^2 + \frac{(C'')^2}{\delta}|u|_{m-1}^2 \leqq 2\delta|u|_m^2 + C(\delta)|u|_0^2$$

by proposition 3.5.16. If $\delta \leqq \frac{1}{2}c$, these give us

$$\tfrac{1}{2}c|u|_m^2 \leqq (-1)^m \operatorname{Re} \langle Pu, u\rangle + B_0|u|_0^2$$

for a suitable $B_0 > 0$ and all $u \in C_0^\infty(\Omega', r)$.

3.6.6 Proposition. If P is a uniformly strongly elliptic operator of order $2m$ which is homogeneous and has constant coefficients, i.e.

$$Pu(x) = \sum_{|\alpha|=2m} a_\alpha D^\alpha u(x),$$

then, the above inequality can be sharpened a little. One now has, for any $\Omega \Subset R^n$,

$$(-1)^m \operatorname{Re} \langle Pu, u\rangle \geqq c(\Omega)|u|_m^2$$

for all $u \in C_0^\infty(\Omega, r)$.

Proof. We have

$$(-1)^m\langle Pu, u\rangle = \int_{R^n} (p(\xi)\hat{u}(\xi), \hat{u}(\xi))d\xi \geqq c\int_{R^n} |\xi|^{2m}|\hat{u}(\xi)|^2 d\xi,$$

our assertion follows from inequality 3.4.5 and remark 3.5.7.

3.6.7 Proposition. Let Ω be an open set in R^n and let $\varphi \in C_0^\infty(\Omega)$ and $k \geqq 1$ be an integer. Then for any $\varepsilon > 0$, there is $C(\varepsilon) > 0$ such that, for all $f \in C^\infty(\Omega)$, we have

$$\sum_{|\alpha|=k} |\varphi^k D^\alpha f|_0^2 \leqq \varepsilon \sum_{|\alpha|=k+1} |\varphi^{k+1} D^\alpha f|_0^2 + C(\varepsilon) \sum_{|\alpha|=k-1} |\varphi^{k-1} D^\alpha f|_0^2,$$

(φ^0 stands for the function 1).

Proof. It suffices to show that for $k \geqq 1$, $|\beta| = k$, there is a $C(\varepsilon)$ with

$$|\varphi^k D^\beta f|_0^2 \leqq \varepsilon \sum_{|\alpha|=k+1} |\varphi^{k+1} D^\alpha f|_0^2 + C(\varepsilon) \sum_{|\alpha|=k-1} |\varphi^{k-1} D^\alpha f|_0^2.$$

Now, we write $\beta = \beta' + \gamma$, where $|\beta'| = k-1$, $|\gamma| = 1$. Then

$$\begin{aligned} |\varphi^k D^\beta f|_0^2 &= \langle D^\beta f, \varphi^{2k} D^\beta f\rangle = -\langle D^{\beta'} f, D^\gamma(\varphi^{2k} D^\beta f)\rangle \\ &= -\langle D^{\beta'} f, 2k\varphi^{2k-1} D^\gamma \varphi D^\beta f\rangle - \langle D^{\beta'} f, \varphi^{2k} D^{\beta+\gamma} f\rangle \\ &= -2k\langle D^\gamma \varphi \cdot \varphi^{k-1} D^{\beta'} f, \varphi^k D^\beta f\rangle - \langle \varphi^{k-1} D^{\beta'} f, \varphi^{k+1} D^{\beta+\gamma} f\rangle. \end{aligned}$$

Using the fact that

$$2|\langle u, v\rangle| \leqq \delta |u|_0^2 + \frac{1}{\delta}|v|_0^2 \qquad \text{for all} \quad \delta > 0,$$

we obtain, for any $\delta > 0$,

$$|\varphi^k D^\beta f|_0^2 \leqq \delta\{|\varphi^k D^\beta f|_0^2 + |\varphi^{k+1} D^{\beta+\gamma} f|_0^2\} + C_1(\delta)|\varphi^{k-1} D^{\beta'} f|_0^2,$$

and the required inequality follows easily.

3.6.8 FRIEDRICHS' INEQUALITY. Let Ω be a bounded open set in R^n and P a linear elliptic differential operator from ϑ_r to ϑ_s of order m given by

$$(Pu)(x) = \sum_{|\alpha| \leqq m} a_\alpha(x)(D^\alpha u)(x), \qquad u \in C^\infty(\Omega, r).$$

Let k be an integer $\geqq 0$.

3.6.9 If the a_α's are constant, then there is a constant $C > 0$ such that

$$|u|_{m+k} \leqq C|Pu|_k$$

for all $u \in C_0^\infty(\Omega, r)$.

3.6.10 For any $x_0 \in \Omega$, there is a neighbourhood U of x_0 and a constant $C_1 > 0$ such that

$$|u|_{m+k} \leqq C_1|Pu|_k$$

for all $u \in C_0^\infty(U, r)$ (here P does not need to have constant coefficients).

3.6.11 If Ω' is relatively compact in Ω, there exists a constant $C_2 > 0$ such that

$$|u|_{m+k}^{\Omega'} \leqq C_2\{|Pu|_k^\Omega + |u|_0^\Omega\}$$

for all $u \in C^\infty(\Omega, r)$.

In particular,

$$|u|_{m+k} \leqq C_2\{|Pu|_k + |u|_0\}$$

for all $u \in C_0^\infty(\Omega', r)$.

PROOF OF 3.6.9. Let $p(x, \xi) = p(\xi)$ be the characteristic polynomial of P. Since P is elliptic, $p(\xi)\, v \neq 0$ for $|\xi| = 1$, $v \in C^r$, $|v| = 1$. Hence, by homogeneity, there is $\rho_1 > 0$ so that

$$\|p(\xi)v\| \geqq \rho_1|\xi|^m|v|, \qquad \xi \in R^n, v \in C^r.$$

We have

$$|Pu|_k^2 \geqq C' \int_{R^n} (1+|\xi|^2)^k |\widehat{Pu}(\xi)|^2 \, d\xi \geqq C' \int_{|\xi| \geqq M} (1+|\xi|^2)^k |\widehat{Pu}(\xi)|^2 \, d\xi,$$

for any $M > 0$.
Now

$$\widehat{Pu}(\xi) = i^{m} p(\xi)\hat{u}(\xi) + q(\xi)\hat{u}(\xi),$$

where q is a polynomial of degree $\leqq m-1$. Since, for complex numbers a, b we have

$$|a+b|^2 \geqq \tfrac{1}{2}|a|^2 - |b|^2,$$

it follows that there is a constant $A > 0$ so that

$$\begin{aligned} |\widehat{Pu}(\xi)|^2 &\geqq \tfrac{1}{2}|p(\xi)\hat{u}(\xi)|^2 - A(1+|\xi|^2)^{m-1}|\hat{u}(\xi)|^2 \\ &\geqq \tfrac{1}{2}\rho_1|\xi|^{2m}|\hat{u}(\xi)|^2 - A(1+|\xi|^2)^{m-1}|\hat{u}(\xi)|^2 \\ &\geqq \rho_2(1+|\xi|^2)^m|\hat{u}(\xi)|^2 \qquad \text{if} \quad |\xi| \geqq M \end{aligned}$$

where M and $\rho_2 > 0$ are suitably chosen. With such a choice of M, we have

$$|Pu|_k^2 \geqq C'\rho_2 \int_{|\xi| \geqq M} (1+|\xi|^2)^{m+k} |\hat{u}(\xi)|^2 \, d\xi$$

and (3.6.9) follows from propositions 3.4.7 and 3.5.6.

Proof of 3.6.10. Write P in the form $P = P^0 + P'$ where P^0 has constant coefficients and all the coefficients of P' vanish at x_0 (and P^0 and P' have order m). We have only to show that if $\varepsilon > 0$ is given, there is a neighbourhood U of x_0 so that

$$|P'u|_k \leqq \varepsilon |u|_{m+k}$$

for $u \in C_0^\infty(U, r)$.
Consider

$$D^\beta(bD^\alpha f), \quad |\alpha| \leqq m, \ |\beta| \leqq k, \ b \in C^\infty(\Omega), f \in C_0^\infty(\Omega), \quad b(x_0)=0.$$

We have

$$D^\beta(bD^\alpha f) = \sum_{\gamma \leqq \beta} \binom{\beta}{\gamma} D^\gamma b D^{\alpha-\gamma} f.$$

One sees immediately, by integrating by parts, that

$$|D^\gamma b D^{\alpha-\gamma} f|_0^2 \leqq \text{const } ||b||_{2k} \cdot \int_U |b(x)| \sum_{|\lambda| \leqq m+k} |D^\lambda f(x)|^2 \mathrm{d}x.$$

where

$$||b||_{2k} = \sum_{|\lambda| \leqq 2k} \sup_U |D^\lambda b(x)|.$$

Since $b(x_0) = 0$, it follows that if U is small enough, then

$$|D^\beta(bD^\alpha f)|_0^2 \leqq \varepsilon |f|_{m+k}^2$$

for $f \in C_0^\infty(U)$, $|\alpha| \leqq m$, $|\beta| \leqq k$.
Hence, if U is a small enough neighbourhood of x_0, we have

$$|P'u|_k \leqq \varepsilon |u|_{m+k}$$

for $u \in C_0^\infty(U, r)$.

PROOF OF 3.6.11. We first remark that

$$|u|_{m+k} \leqq C'\{|Pu|_k + |u|_0\}$$

for all $u \in C_0^\infty(\Omega', r)$.
In fact, let $\{U_1, \ldots, U_h\}$ be a covering of $\bar{\Omega}'$ so that

$$|u|_{m+k} \leqq C|Pu|_k$$

for all $u \in C_0^\infty(U_j, r)$.
Let

$$\eta_j \in C_0^\infty(U_j), \qquad 0 \leqq \eta_j \leqq 1, \qquad \sum \eta_j^2(x) = 1$$

for all x in a neighbourhood of $\bar{\Omega}'$. We have

$$\left|\, |u|_{m+k}^2 - \sum_j |\eta_j u|_{m+k}^2 \right| \leqq \text{const } |u|_{m+k-1}^2,$$

$$\left|\, |Pu|_k^2 - \sum_j |P(\eta_j u)|_k^2 \right| \leqq \text{const } |u|_{m+k-1}^2.$$

This gives, since

$$|\eta_j u|_{m+k} \leqq \text{const } |P(\eta_j u)|_k,$$

the inequality

$$|u|_{m+k}^2 \leqq \text{const } \{|Pu|_k^2 + |u|_{m+k-1}^2\}$$

and the remark follows from proposition 3.5.16.
Now let

$$\varphi \in C_0^\infty(\Omega), \qquad 0 \leqq \varphi \leqq 1, \qquad \varphi(x) = 1$$

for all x in a neighbourhood of $\bar{\Omega}'$, and let $u \in C^\infty(\Omega, r)$. Let $m' = m+k$. We have for $|\alpha| \leqq m'$,

$$D^\alpha(\varphi^{m'}u) = \sum_{\beta \leqq \alpha} \binom{\alpha}{\beta} D^\beta \varphi^{m'} D^{\alpha-\beta} u$$
$$= \varphi^{m'} D^\alpha u + \sum_{\beta \leqq \alpha, \beta \neq 0} c_\beta \varphi^{m'-|\beta|} D^{\alpha-\beta} u, \qquad c_\beta \in C_0^\infty(\Omega),$$

Squaring both sides, using Schwarz' inequality and summing over all α with $|\alpha| \leqq m' = m+k$, we obtain

$$\Big| |\varphi^{m+k}u|^2_{m+k} - \sum_{|\alpha| \leqq m+k} |\varphi^{m+k} D^\alpha u|^2_0 \Big| \leqq \text{const} \sum_{|\beta| < m+k} |\varphi^{|\beta|} D^\beta u|^2_0 .$$

Now, by the result we have established above, we have

$$|\varphi^{m+k}u|^2_{m+k} \leqq \text{const}\, \{|P(\varphi^{m+k}u)|^2_k + |\varphi^{m+k}u|^2_0\}.$$

Furthermore, it follows from proposition 3.6.7 that we have

$$\sum_{|\beta| < m+k} |\varphi^{|\beta|} D^\beta u|^2_0 \leqq \varepsilon \sum_{|\beta| = m+k} |\varphi^{m+k} D^\beta u|^2_0 + C(\varepsilon)(|u|^\Omega_0)^2.$$

Putting these together, and choosing ε small enough, we deduce that

$$\sum_{|\alpha| \leqq m+k} |\varphi^{m+k} D^\alpha u|^2_0 \leqq \text{const}\, \{|\varphi^{m+k} P u|^2_k + (|u|^\Omega_0)^2\}.$$

Since $\varphi(x) = 1$ for $x \in \Omega'$ and supp $(\varphi) \subset \Omega$, (3.6.11) follows from this inequality.

An immediate corollary of (3.6.11) is the following theorem:

3.6.12 Theorem. Let P be an elliptic differential operator from ϑ_r to ϑ_s on the open set $\Omega \subset R^n$. Let

$$\mathscr{P} = \{u \in C^\infty(\Omega, r) | Pu = 0\}.$$

Then, a sequence $\{u_\nu\}$ of elements in $\mathscr{P}$ converges in the topology of $C^\infty(\Omega, r)$ if and only if, for any compact subset $K \subset \Omega$,

$$\int_K |u_\nu - u_\mu|^2 \mathrm{d}x \to 0 \quad \text{as} \quad \nu, \mu \to \infty.$$

More generally, if $\{u_\nu\}$ is a sequence of elements in $C^\infty(\Omega, r)$ such that $\{Pu_\nu\}$ converges in $C^\infty(\Omega, s)$ and

$$\int_K |u_\nu - u_\mu|^2 \mathrm{d}x \to 0$$

for any compact $K \subset \Omega$, then $\{u_\nu\}$ converges in $C^\infty(\Omega, r)$.

The proofs given in this section are essentially those of GÅRDING [1953] and FRIEDRICHS [1953].

§ 3.7 Elliptic operators with C^∞ coefficients: the regularity theorem

Let Ω be an open set in R^n and P a linear differential operator from ϑ_r to ϑ_s on Ω. Let $u \in H_0(\Omega)$. We define Pu as a linear functional on $C_0^\infty(\Omega, s)$ defined by

$$(Pu)(v) = \langle u, P^*v\rangle, \qquad v \in C_0^\infty(\Omega, s).$$

If for some q, $1 \leqq q \leqq \infty$, there is a $g \in L^q(\Omega, s)$ with

$$(Pu)(v) = \langle g, v\rangle$$

for all $v \in C_0^\infty(\Omega, s)$, then, we identify the linear functional Pu with g and write $Pu = g$, $Pu \in L^q(\Omega, s)$. Note that this is consistent with the usual notation if $u \in C^\infty(\Omega, r)$. Further, if P has order m and $u \in H_m(\Omega)$, then $Pu \in H_0(\Omega)$. If g belongs to some subspace of $L^2(\Omega, s) = H_0(\Omega)$, we shall also say that Pu lies in this subspace.

Let P be a uniformly strongly elliptic operator from ϑ_r to itself of order $2m$, and write

$$P = \sum_{\nu=1}^{N} R_\nu^* \circ Q_\nu,$$

where Q_ν, R_ν are operators of order $\leqq m$ from ϑ_r to itself. For $u, v \in C_0^\infty(\Omega, r)$, set

$$H\langle u, v\rangle = \sum_{\nu=1}^{N} \langle Q_\nu u, R_\nu v\rangle.$$

Clearly, this definition extends to elements $u, v \in H_m(\Omega)$. Let h be a real number, different from zero. Let $\Omega' \Subset \Omega$, and let $x = (x_1, \ldots, x_n) \in \Omega'$. We write

$$x+h = (x_1+h, x_2, \ldots, x_n)$$

and suppose that, for $x \in \Omega'$, we have $x+h \in \Omega$. If $g \in H_m(\Omega)$, we set

$$g^h(x) = h^{-1}\{g(x+h)-g(x)\}.$$

These conventions apply for propositions 3.7.1–3.7.4.

3.7.1 PROPOSITION. If $\eta \in C_0^\infty(\Omega, 1)$, there is a constant $C > 0$ such that for all $f \in H_0(\Omega)$, we have

$$|(\eta f)^h - \eta(f^h)|_0 \leqq C|f|_0 .$$

Further, there is a $C > 0$ such that for any $u \in \mathring{H}_m(\Omega)$, we have $|u^h|_{m-1} \leqq C|u|_m$ for small enough h.

PROOF. We have

$$(\eta f)^h(x) - (\eta f^h)(x) = \eta^h(x) f(x+h),$$

and the first inequality follows. As for the second, we have

$$u^h(x) = \int_0^1 \frac{\partial u}{\partial x_1}(x_1+th, x_2, \ldots, x_n)\mathrm{d}t$$

if $u \in C_0^\infty(\Omega, r)$, so that the required inequality follows if $u \in C_0^\infty(\Omega, r)$. The general inequality then follows by closure.

3.7.2 THEOREM. Let $f \in H_m(\Omega)$, $m \geqq 1$, have compact support in Ω. Suppose that there is a $C > 0$ such that

$$|H\langle f, u\rangle| \leqq C|u|_{m-1}$$

for all $u \in C_0^\infty(\Omega, r)$. Then $f \in \mathring{H}_{m+1}(\Omega)$.

PROOF. Let $h \neq 0$ be sufficiently small. We shall write $\mathrm{O}(|u|_m)$ for any complex valued function G of u and h which satisfies

$$|G| \leqq \text{const } |u|_m ,$$

where the constant may depend on f but not on u or h. We have

$$H\langle f^h, u\rangle = \sum_{\nu=1}^{N} \langle Q_\nu f^h, R_\nu u\rangle.$$

Since $D^\alpha(f^h) = (D^\alpha f)^h$, it follows from proposition 3.7.1, that $|Q_\nu f^h - (Q_\nu f)^h|_0$ is bounded as a function of h. Hence

$$H\langle f^h, u\rangle = \sum_\nu \langle (Q_\nu f)^h, R_\nu u\rangle + O(|u|_m).$$

On the other hand, we have

$$\begin{aligned}\langle (Q_\nu f)^h, R_\nu u\rangle &= -\langle Q_\nu f, (R_\nu u)^{-h}\rangle \\ &= -\langle Q_\nu f, R_\nu u^{-h}\rangle + O(|u|_m),\end{aligned}$$

so that

$$H\langle f^h, u\rangle = -H\langle f, u^{-h}\rangle + O(|u|_m).$$

Now, by hypothesis, we have

$$|H\langle f, u^{-h}\rangle| \leqq C|u^{-h}|_{m-1} \leqq C'|u|_m$$

(proposition 3.7.1).
Hence, there is a constant $C_1 > 0$ so that

$$|H\langle f^h, u\rangle| \leqq C_1|u|_m \qquad \text{for all} \quad u \in C_0^\infty(\Omega, r).$$

Let $\{u_\nu\}$ be a sequence of elements in $C_0^\infty(\Omega, r)$ converging in $H_m(\Omega)$ to f^h. We have

$$|H\langle f^h, f^h\rangle| = \lim |H\langle f^h, u_\nu\rangle| \leqq C_1 \lim |u_\nu|_m = C_1|f^h|_m.$$

Also, by inequality 3.6.3, there is a $C_2 > 0$ so that

$$|u_\nu|_m^2 \leqq C_2\{|\langle Pu_\nu, u_\nu\rangle| + |u_\nu|_0^2\} = C_2\{|H\langle u_\nu, u_\nu\rangle| + |u_\nu|_0^2\},$$

so that, letting $\nu \to \infty$, we have

$$|f^h|_m^2 \leqq C_2\{|H\langle f^h, f^h\rangle| + |f^h|_0^2\} \leqq C_3\{|f^h|_m + |f_0^h|_0^2\}.$$

Since $m \geqq 1$, by proposition 3.7.1, $|f^h|_0$ is bounded as $h \to 0$. Hence, as $h \to 0$, we have

$$|f^h|_m^2 \leqq C_3|f^h|_m + C_4.$$

This implies that $|f^h|_m$ is bounded as $h \to 0$. Thus $\{f^h\}$ is a bounded set in the Hilbert space $H_m(\Omega)$. Hence, there is a sequence $\{h_\mu\}$, $h_\mu \to 0$, so that f^{h_μ} converges weakly to an element g in $H_m(\Omega)$, i.e.

$$[f^{h_\mu}, v]_m \to [g, v]_m$$

for all $v \in H_m(\Omega)$ (see equation 3.4.6).
Moreover, since $m \geqq 1$,

$$f^h \to \frac{\partial f}{\partial x_1}$$

in $H_0(\Omega)$. It follows that $\partial f/\partial x_1 = g \in H_m(\Omega)$. In exactly the same way, we see that $\partial f/\partial x_j \in H_m(\Omega)$ for $j = 1, \ldots, n$. Hence, by theorem 3.4..12, f is $(m+1)$ times strongly differentiable. Since f has compact support, $f \in \mathring{H}_{m+1}(\Omega)$.

3.7.3 PROPOSITION. Let $f \in H_k(\Omega)$ and L be a linear differential operator on a neighbourhood of $\bar{\Omega}$ from ϑ_r to itself. If order $L \leqq p, p \geqq k$, then, for all $u \in C_0^\infty(\Omega, r)$ we have

$$|\langle f, Lu \rangle| \leqq \text{const } |u|_{p-k}.$$

PROOF. If we write $L = \sum B_\nu^* \circ A_\nu$ where order $A_\nu \leqq p-k$, order $B_\nu \leqq k$, we have

$$\langle f, Lu \rangle = \sum \langle B_\nu f, A_\nu u \rangle$$

and the result follows since $B_\nu f \in H_0(\Omega)$.

3.7.4 PROPOSITION. Let $f \in H_m(\Omega)$ and, for an integer μ, $0 < \mu \leqq m$, suppose that there is a constant $C > 0$ so that

$$|H\langle f, u \rangle| \leqq C|u|_{m-\mu} \qquad \text{for all} \quad u \in C_0^\infty(\Omega, r).$$

Then f is $(m+\mu)$ times strongly differentiable.

PROOF. We proceed by induction on μ.
Case $\mu = 1$: Suppose that

$$|H\langle f, u \rangle| \leqq C|u|_{m-1}.$$

Let $\eta \in C_0^\infty(\Omega, 1)$. We have, with $P = \sum R_\nu^* \circ Q_\nu$ as before,

$$\sum \langle \eta Q_\nu f, R_\nu u \rangle - \sum \langle Q_\nu(\eta f), R_\nu u \rangle = \langle f, L'u \rangle,$$

where order $L \leqq 2m-1$(since $\eta Q_\nu f - Q_\nu(\eta f) = Qf$ is a linear differential operator of order $\leqq m-1$). Further

$$\sum \langle \eta Q_\nu f, R_\nu u \rangle - \sum \langle Q_\nu f, R_\nu(\bar{\eta} f) \rangle = \langle f, L'u \rangle,$$

where L' again has order $\leqq 2m-1$. Hence

$$H\langle \eta f, u\rangle - H\langle f, \bar{\eta}u\rangle = \langle f, Lu\rangle,$$

where L has order $\leqq 2m-1$. Hence, since $f \in H_m(\Omega)$, proposition 3.7.3 implies that

$$|H\langle \eta f, u\rangle - H\langle f, \bar{\eta}u\rangle| \leqq \text{const } |u|_{m-1}.$$

Since, by assumption, $|H\langle f, \bar{\eta}u\rangle| \leqq \text{const } |\bar{\eta}u|_{m-1}$, this gives

$$|H\langle \eta f, u\rangle| \leqq \text{const } |u|_{m-1},$$

so that, by proposition 3.7.1, $\eta f \in \mathring{H}_{m+1}(\Omega)$. Since η is arbitrary, this implies that f is $(m+1)$ times strongly differentiable.

General case: Suppose now the result proved for $\mu = \mu_0 - 1$, $1 < \mu_0 \leqq m$, and that

$$|H\langle f, u\rangle| \leqq \text{const } |u|_{m-\mu_0}.$$

By induction, f is $(m+\mu_0-1)$ times strongly differentiable. By restricting our attention to an open set $\Omega' \Subset \Omega$, we may suppose that $f \in H_{m+\mu_0-1}(\Omega)$. Let $\alpha = (\alpha_1, \ldots, \alpha_n)$ be an n-tuple with $|\alpha| = 1$. We see at once that

$$H\langle D^\alpha f, u\rangle + H\langle f, D^\alpha u\rangle = \langle f, Lu\rangle,$$

where L is a differential operator of order $\leqq 2m$. Again, by proposition 3.7.3, this shows, since $f \in H_{m+\mu_0-1}(\Omega)$, that

$$|H\langle D^\alpha f, u\rangle + H\langle f, D^\alpha u\rangle| \leqq \text{const } |u|_{m-\mu_0+1}.$$

Since

$$|H\langle f, D^\alpha u\rangle| \leqq \text{const } |D^\alpha u|_{m-\mu_0} \leqq \text{const } |u|_{m-\mu_0+1},$$

this shows that

$$|H\langle D^\alpha f, u\rangle| \leqq \text{const } |u|_{m-\mu_0+1} = \text{const } |u|_{m-\mu}.$$

By induction hypothesis, this implies that $D^\alpha f$ is $(m+\mu_0-1)$ times strongly differentiable. By theorem 3.4.12, our proposition follows.

3.7.5 PROPOSITION. If $f \in H_m(\Omega)$ and Pf is μ times strongly differentiable, $\mu \geqq 0$, then f is $(2m+\mu)$ times strongly differentiable.

PROOF. We first prove the theorem for $\mu = 0$. If $Pf \in H_0(\Omega)$, this means that there is $g \in H_0(\Omega)$ with

$$\langle f, P^*u\rangle = \langle g, u\rangle \qquad \text{for all} \quad u \in C_0^\infty(\Omega, r).$$

This implies, since $f \in H_m(\Omega)$, that

$$H\langle f, u\rangle = \langle g, u\rangle,$$

so that $|H\langle f, u\rangle| \leqq \text{const}\, |u|_0$. By proposition 3.7.4, f is $2m$ times strongly differentiable.

Suppose now that it has already been shown that f is $(2m+\mu-1)$ times strongly differentiable. By replacing Ω by a relatively compact open subset of Ω, we may suppose that $f \in H_{2m+\mu-1}(\Omega)$, and that there is a $g \in H_\mu(\Omega)$ with

$$\langle f, P^*u\rangle = \langle g, u\rangle, \qquad u \in C_0^\infty(\Omega, r).$$

Now, if $|\alpha| \leqq \mu$, consider the operator

$$L = P \circ D^\alpha - D^\alpha \circ P,$$

where D^α is the usual operator of derivation from ϑ_r to itself given by

$$(u_1, \ldots, u_r) \mapsto (D^\alpha u_1, \ldots, D^\alpha u_r).$$

Clearly L has order $\leqq 2\,m+\mu-1$. Hence

$$P(D^\alpha f) = D^\alpha g + Lf,$$

where $g = Pf \in H_\mu(\Omega)$. Since $f \in H_{2m+\mu-1}(\Omega)$ by induction hypothesis, this implies that

$$P(D^\alpha f) \in H_0(\Omega).$$

By the special case proved above, $D^\alpha f$ is $2m$ times strongly differentiable. Since α, $|\alpha| \leqq \mu$ was arbitrary, the proposition is proved.

3.7.6 Proposition. Let Δ be the Laplace operator from ϑ_r to ϑ_r on R^n (example 3.3.15, (c)), If $f \in H_0(R^n)$ and $p \geqq 1$ is an integer, there is an $F \in H_{2p}(R^n)$ such that $(I-\Delta)^p F = f$; here $I-\Delta$ is the operator

$$u \mapsto u - \Delta u$$

and

$$(I-\Delta)^p = (I-\Delta) \circ \ldots \circ (I-\Delta),$$

where product is taken p times.

PROOF. By Plancherel's theorem 3.2.8, $\hat{f} \in L^2(R^n)$. Let

$$\psi(\xi) = \hat{f}(\xi)(1+\xi_1^2+\ldots+\xi_n^2)^{-\rho}.$$

Since $\psi \in L^2(R^n)$, there is an $F \in L^2(R^n)$ with $\hat{F} = \psi$. Moreover

$$\int_{R^n} (1+|\xi|^2)^{2\rho} |\hat{F}(\xi)|^2 \, d\xi < \infty.$$

Hence, by (3.4.9), $F \in H_{2\rho}(R^n)$. We see at once that

$$(I-\Delta)^\rho F = f.$$

We now consider an arbitrary elliptic operator P from ϑ_r to ϑ_s on an open set $\Omega \subset R^n$.

3.7.7 REGULARITY THEOREM. Let P be an elliptic differential operator from ϑ_r to ϑ_s on the open set $\Omega \subset R^n$. Suppose that $f \in H_0(\Omega)$ and $Pf \in C^\infty(\Omega, s)$. Then $f \in C^\infty(\Omega, r)$.

PROOF. Consider the operator

$$L = (-1)^m P^* \circ P, \qquad m = \text{order of } P.$$

Then L is an elliptic operator from ϑ_r to ϑ_s, and, if $\Omega' \Subset \Omega$, then L is uniformly strongly elliptic on Ω' (corollary 3.6.2). Further

$$Lf = (-1)^m P^* g, \qquad g = Pf,$$

so that, if $Pf \in C^\infty(\Omega, s)$, then $Lf \in C^\infty(\Omega, r)$. Hence, we may suppose that P is uniformly strongly elliptic from ϑ_r to ϑ_r of order $2m$.

If $f \in H_0(\Omega)$, define $f_0 \in H_0(R^n)$ by

$$f_0(x) = \begin{cases} f(x) & \text{if } x \in \Omega, \\ 0 & \text{if } x \notin \Omega. \end{cases}$$

By proposition 3.7.6, there is an $F_0 \in H_{2m}(R^m)$ such that

$$(I-\Delta)^m F_0 = f_0.$$

Let

$$P_0 = (-1)^m P \circ (I-\Delta)^m.$$

Then P_0 is again a uniformly strongly elliptic operator of order $4m$.

Moreover, if $F_0|\Omega = F$, we have

$$F \in H_{2m}(\Omega), \qquad P_0 F = (-1)^m Pf \in C^\infty(\Omega, r).$$

Hence, by proposition 3.7.4, F is $(4m+\mu)$ times strongly differentiable for any $\mu \geqq 0$. By proposition 3.5.13, $F \in C^\infty(\Omega, r)$. Hence

$$f = (-1)^m (I - \Delta)^m F \in C^\infty(\Omega, r).$$

The proof of the regularity theorem given here is essentially that of Nirenberg [1955]. There are now several other proofs available. The oldest, which operates with 'fundamental solutions' was proposed by L. Schwartz [1950/51]. Very strong theorems that can be obtained by this method will be found in Hörmander [1963]. The first proof using only *a priori* estimates (inequalities of the Gårding, Friedrichs type) is due to Friedrichs [1953] (who proves however only a somewhat weaker result). Other proofs are due to John [1955] and Lax [1955]. The proof of Lax is both brief and elegant, but uses also the duals of the spaces $H_m(\Omega)$. Another very general and elegant proof, using the so called pseudo-differential operators, will be found in L. Schwartz [1963/64]. See also Hörmander [1965] and Nirenberg [1970]. There is a vast literature that has sprung up around this theorem and its generalizations, for instance to the problem of 'regularity at the boundary' or corresponding results for parabolic and other operators.

§ 3.8 Elliptic operators with analytic coefficients

3.8.1 Proposition. If K is a compact set in R^n, set

$$K_\delta = \{x \in R^n |\, d(x, K) < \delta\}.$$

There are constants $C_\alpha > 0$ such that given a compact set $K \subset R^n$ and $\delta > 0$, there exists

$$\varphi_\delta \in C_0^\infty(R^n), \qquad 0 \leqq \varphi_\delta \leqq 1, \qquad \varphi_\delta(x) = 1 \text{ for } x \in K,$$

$$\operatorname{supp}(\varphi_\delta) \subset K_\delta \quad \text{and} \quad |D^\alpha \varphi_\delta(x)| \leqq C_\alpha \delta^{-|\alpha|}$$

for all $\alpha = (\alpha_1, \ldots, \alpha_n)$.

PROOF. Let ψ be a C^∞ function with

$$\int_{R^n} \psi(x)dx = 1, \qquad \psi \geqq 0 \quad \text{and} \quad \operatorname{supp}(\psi) \subset \{x \mid \|x\| < 1\}.$$

Let χ_δ be the characteristic function of K_δ, i.e.

$$\chi_\delta(x) = \begin{cases} 1 & \text{if } x \in K_\delta, \\ 0 & \text{if } x \notin K_\delta. \end{cases}$$

Let

$$\psi_\delta(x) = \delta^{-n} \int_{R^n} \psi((x-y)/\delta)\chi_\delta(y)dy.$$

Then $\psi_\delta(x) = 1$ for $x \in K$ and $\operatorname{supp}(\psi_\delta) \subset K_{2\delta}$. Moreover

$$D^\alpha \psi_\delta(x) = \delta^{-n-|\alpha|} \int_{R^n} (D^\alpha \psi)((x-y)/\delta)\chi_\delta(y)dy$$

so that

$$|D^\alpha \psi_\delta(x)| \leqq \delta^{-|\alpha|} \int_{R^n} |D^\alpha \psi(y)| dy.$$

We have only to take $\varphi_\delta = \psi_{\delta/2}$.

3.8.2 REMARK. In what follows, R and ρ are real numbers such that $0 < \rho < \min\{1, R\}$, and we set, $\Omega_0 = \{x \mid \|x\| < R\}$. For $f \in L^2(\Omega, r)$, let

$$M_\rho(f)^2 = \int_{\|x\| < R-\rho} |f(x)|^2 dx.$$

Let $\delta > 0$ and $\Omega = \{x \mid \|x\| < R+\delta\}$.

3.8.3 PROPOSITION. Let P be an elliptic operator of order m from ϑ_r to itself on Ω. Then there is a constant $C > 0$ such that for all $u \in C^\infty(\Omega, r)$, and positive ρ, ρ' with $\rho + \rho' < R$, we have

$$\rho^m M_{\rho+\rho'}(D^\alpha u) \leqq C\{\rho^m M_{\rho'}(Pu) + \sum_{|\beta|<m} \rho^{|\beta|} M_{\rho'}(D^\beta u)\} \quad \text{for} \quad |\alpha| = m.$$

PROOF. Let $\varphi \in C_0^\infty(R^n)$ be so that $\varphi(x) = 1$ for

$$\|x\| < R-\rho-\rho', \qquad 0 \leqq \varphi \leqq 1$$

and

$$\operatorname{supp}(\varphi) \subset \{x \mid \|x\| < R-\rho'\};$$

further φ can be so chosen that

$$|D^\alpha \varphi| \leqq C_\alpha \rho^{-|\alpha|},$$

the C_α being constants depending only on α and n (proposition 3.8.1). By (3.6.11), there is a constant $A > 0$ so that, for $|\alpha| \leqq m$,

$$|D^\alpha(\varphi u)|_0 \leqq A\{|P(\varphi u)|_0 + |\varphi u|_0\}.$$

Let

$$Pu(x) = \sum_{|\lambda| \leqq m} a_\lambda(x) D^\lambda u(x),$$

where the a_λ are C^∞ on Ω. We have

$$P(\varphi u) - \varphi P(u) = \sum_{\beta < \lambda, |\lambda| \leqq m} a_\lambda \binom{\lambda}{\beta} D^{\lambda-\beta}\varphi D^\beta u.$$

Clearly, there are constants $c_{\lambda,\beta}$ independent of ρ so that

$$\left| a_\lambda(x) \binom{\lambda}{\beta} D^{\lambda-\beta}\varphi(x) \right| \leqq c_{\lambda,\beta}\rho^{-|\lambda-\beta|} \qquad \text{for} \quad \|x\| \leqq R.$$

Hence

$$|D^\alpha(\varphi u)|_0 \leqq \text{const}\,\{M_{\rho'}(Pu) + \sum_{|\beta| < m} \rho^{-m+|\beta|} M_{\rho'}(D^\beta u)\}$$

which completes the proof.

3.8.4 THEOREM. Let P be an elliptic operator of order m from ϑ_r to itself on an open set U in R^n, and suppose that P has analytic coefficients, and that $0 \in U$. If then $R > 0$ and $\delta > 0$ are chosen small enough, there exists a constant $A \geqq 1$ such that for all ρ, $0 < \rho < \min\{1, R\}$ and $u \in C^\infty(\Omega, r)$, we have

3.8.5 $$\rho^{|\alpha|} M_{|\alpha|\rho}(D^\alpha u) \leqq A^{|\alpha|+1} \left\{ \sum_{\nu=1}^{k} \rho^{(\nu-1)m} M(P^\nu u) + M(u) \right\}$$

for $|\alpha| < km$, $k = 1, 2, \ldots$. Here we have set

$$P^\nu = P \circ \ldots \circ P, \quad \text{(product } \nu \text{ times)}$$

and

$$M(f)^2 = M_{-\delta}(f)^2 = \int_{||x|| \leqq R+\delta} |f(x)|^2 dx.$$

PROOF. If

$$(Pu)(x) = \sum_{|\lambda| \leqq m} a_\lambda(x) D^\lambda u(x),$$

the a_λ are real analytic on U. Hence, if $R_1 > 0$ is small enough, then the a_λ are the restrictions to $||x|| \leqq R_1$ of holomorphic mappings, which we denote again by a_λ, of $\{z \in C^n | \; ||z|| \leqq R_1\}$ into the space of $r \times r$ complex matrices.

Let

$$B = \sum_{|\lambda| \leqq m} \sup_{||z|| \leqq R_1} |a_\lambda(z)|.$$

By Cauchy's inequalities 1.1.4 we have

3.8.6 $$\sum_{|\lambda| \leqq m} |D^\alpha a_\lambda(x)| \leqq B\alpha! \rho^{-|\alpha|} \quad \text{for} \quad ||x|| \leqq R_1 - \rho.$$

Let $0 < R < R_1$, $\delta = R_1 - R$, and set

$$S_k(u) = S_k(u, \rho) = \sum_{\nu=1}^{k} \rho^{(\nu-1)m} M(P^\nu u) + M(u).$$

Then we have

3.8.7 $$\rho^m S_k(Pu) \leqq S_{k+1}(u).$$

We shall prove the inequality (3.8.5) by induction on k. For $k = 1$, i.e. for $|\alpha| \leqq m$, we have, by (3.6.11),

$$M_0(D^\alpha u) \leqq C_2\{M(Pu) + M(u)\},$$

and (3.8.5) holds (for $\rho \leqq 1$) as soon as $A > C_2$. Suppose now that $km < |\alpha| \leqq (k+1)m$, and suppose that (3.8.5) is proved for all β with $|\beta| < |\alpha|$. Let $\alpha = \alpha_0 + \alpha'$, where $|\alpha_0| = m$. We apply proposition 3.8.3 with $\rho' = (|\alpha| - 1)\rho$, and α_0 in place of α, $D^{\alpha'} u$ in place of u. This gives us

3.8.8 $$\rho^{|\alpha|} M_{|\alpha|\rho}(D^\alpha u) \leqq C\{\rho^{|\alpha|} M_{(|\alpha|-1)\rho}(PD^{\alpha'} u) + \sum_{|\beta| < m} \rho^{|\beta| + |\alpha'|} M_{(|\alpha|-1)\rho}(D^{\beta+\alpha'} u)\}.$$

Further

$$D^{\alpha'}Pu - PD^{\alpha'}u = \sum_{|\lambda| \leqq m} \sum_{\gamma < \alpha'} \binom{\alpha'}{\gamma} D^{\alpha'-\gamma} a_\lambda D^{\gamma+\lambda} u.$$

Now, for $||x|| \leqq R - mk\rho$, we have

$$|D^{\alpha'-\gamma} a_\lambda(x)| \leqq B(\alpha'-\gamma)!(mk\rho)^{-|\alpha'-\gamma|} \text{ (see 3.8.6)},$$

and

$$\binom{\alpha'}{\gamma} \frac{(\alpha'-\gamma)!}{(mk)^{|\alpha'-\gamma|}} \leqq \left(\frac{|\alpha'|}{mk}\right)^{|\alpha'-\gamma|} \leqq 1$$

since $|\alpha'| = |\alpha| - m \leqq km$. Hence, for $||x|| \leqq R - mk\rho$ and *a fortiori* for $||x|| \leqq R - (|\alpha| - 1)\rho$, we have

3.8.8′ $$|D^{\alpha'}Pu(x) - PD^{\alpha'}u(x)| \leqq B \sum_{|\lambda| \leqq m} \sum_{\gamma < \alpha'} \rho^{-|\alpha'-\gamma|} |D^{\gamma+\lambda} u(x)|.$$

This, with (3.8.8) gives us:

For $km < |\alpha| \leqq (k+1)m$, we have

$$\rho^{|\alpha|} M_{|\alpha|\rho}(D^\alpha u) \leqq C\{\rho^{|\alpha|} M_{|\alpha'|\rho}(D^{\alpha'}Pu) + \sum_{|\beta|<m} \rho^{|\beta|+|\alpha'|} M_{|\beta+\alpha'|\rho}(D^{\beta+\alpha'}u)$$
$$+ B \sum_{|\lambda| \leqq m} \sum_{\gamma<\alpha'} \rho^{m+|\gamma|} M_{(m+|\gamma|)\rho}(D^{\gamma+\lambda}u)\}.$$

We can now apply our induction hypothesis to each of the three terms in brackets on the right.

The first term is

$$\leqq \rho^m A^{|\alpha'|+1} S_k(Pu) \leqq A^{|\alpha'|+1} S_{k+1}(u) \quad \text{(see 3.8.7)}.$$

The second term is

$$\leqq \sum_{|\beta|<m} A^{|\beta+\alpha'|+1} S_{k+1}(u)$$

and the third is

$$\leqq B' \sum_{\gamma<\alpha'} A^{m+|\gamma|+1} S_{k+1}(u).$$

This gives

$$\rho^{|\alpha|} M_{|\alpha|\rho}(D^\alpha u) \leqq A^{|\alpha|+1} S_{k+1}(u) \left\{ CA^{-m} + C \sum_{|\beta|<m} \frac{1}{A} + CB' \sum_{\gamma<\alpha'} A^{-|\alpha'-\gamma|} \right\}$$

(since $|\alpha| = m+|\alpha'|$). Now

$$\sum_{\gamma<\alpha'} A^{-|\alpha'-\gamma|} \leqq A^{-1} \sum_{|\beta|\geqq 0} A^{-|\beta|} \to 0 \quad \text{as} \quad A \to \infty.$$

Hence, we can choose $A \geqq C_2$ so large that

$$CA^{-m}+C\sum_{|\beta|<m} \frac{1}{A}+CB'\sum_{\gamma<\alpha'} A^{-|\alpha'-\gamma|} < 1,$$

and this gives us (3.8.5).

3.8.9 THEOREM. (KOTAKÉ–M. S. NARASIMHAN [1962]). Let P be an elliptic operator of order m from ϑ_r to itself on an open set $U \subset R^n$ with analytic coefficients. Let $u \in C^\infty(U, r)$ and suppose that for every $U' \Subset U$, there is an $M > 0$ so that

$$|P^k u|_0^{U'} \leqq M^{k+1}(km)!, \qquad k = 1, 2, \ldots .$$

Then u is analytic in U.

PROOF. We may suppose that $0 \in U$, and it suffices to prove that u is analytic in a neighbourhood of 0. Choose R, δ so that (3.8.5) holds. We then have

$$M(P^\nu u) \leqq M^{\nu+1}(\nu m)!, \qquad \nu = 1, 2, \ldots,$$

so that

$$S_k(u) \leqq \sum_{\nu=1}^{k} M^{\nu+1}\rho^{(\nu-1)m}(\nu m)!+M$$

for any $\rho > 0$.
If we choose $\rho = c/km$, c small, we have

$$(\nu m)!\rho^{(\nu-1)m} \leqq (km)^m$$

for $\nu \leqq k$, so that there is a constant $B_1 > 0$ so that

$$S_k(u) \leqq B_1^{k+1} \qquad \text{for} \quad k \geqq 1.$$

Hence, by (3.8.5), we have

$$M_c(D^\alpha u) \leqq B_2^{k+1}k^k \qquad \text{for} \quad |\alpha| \leqq k.$$

If K is a compact subset of the ball $\{x|\ \|x\| < R-c\}$, we deduce, from proposition 3.5.15 that

$$\sup_{x\in K} |D^\alpha u(x)| \leqq B_3^{k+n+1}(k+n)^{k+n} \qquad \text{for} \quad |\alpha| \leqq k.$$

By Stirling's formula, this implies that

$$\sup_{x\in K} |D^\alpha u(x)| \leqq B_4^{k+1} \cdot k! \qquad \text{for} \quad |\alpha| \leqq k$$

and, by (1.1.15, 16), u is analytic.

3.8.10 THEOREM. Let P be a linear differential operator of order m from ϑ_r to ϑ_r with coefficients which are holomorphic and bounded on

$$D = \{z \in C^n | \ |z_j| < r_j \leqq 1\}.$$

Let u be a bounded holomorphic map: $D \to C^r$. Then there is a constant $A > 0$ (depending on u) such that

$$|P^\nu u(z)| \leqq \frac{(3A)^{\nu+1}(m\nu)!}{\prod_j (r_j - |z_j|)^{m\nu}} \qquad \text{for} \quad z \in D.$$

We first prove:

3.8.11 PROPOSITION. If h is holomorphic in $\{w \in C | \ |w| < R\}$ and

$$|h(w)| < M(R-|w|)^{-\mu}, \qquad |w| < R,$$

then

$$|h'(w)| < 3(\mu+1)M(R-|w|)^{-(\mu+1)}, \qquad |w| < R,$$

where $h'(w)$ is the derivative of h.

PROOF. Let $|w_0| < R$ and $0 < \varepsilon < R-|w_0|$. By Cauchy's inequality

$$|h'(w_0)| \leqq \varepsilon^{-1} \sup_{|w-w_0|=\varepsilon} |h(w)| \leqq M\varepsilon^{-1}(R-|w_0|-\varepsilon)^{-\mu}.$$

If we take

$$\varepsilon = \frac{(R-|w_0|)}{(\mu+1)},$$

we have

$$|h'(\omega_0)| \leqq M(\mu+1)(R-|w_0|)^{-(\mu+1)} \left(\frac{\mu+1}{\mu}\right)^\mu.$$

Since $(1+1/\mu)^\mu < e < 3$, our inequality follows.

PROOF OF THEOREM (3.8.10). We suppose by induction that

$$|P^{\nu-1}u(z)| \leqq \frac{(3A)^{\nu}(m(\nu-1))!}{\prod(r_j-|z_j|)^{m(\nu-1)}}, \qquad z \in D,$$

where

$$\sum_{|\alpha| \leqq m} |a_{\alpha}(z)| \leqq A,$$

and

$$(Pu)(z) = \sum_{|\alpha| \leqq m} a_{\alpha}(z) D^{\alpha} u(z).$$

Then, by proposition 3.8.11, we have

$$|D^{\alpha}P^{\nu-1}u(z)| \leqq \frac{3(3A)^{\nu}(m\nu)!}{\prod(r_j-|z_j|)^{m(\nu-1)+|\alpha|}}$$

and, since

$$\sum_{|\alpha| \leqq m} |a_{\alpha}(z)| \leqq A,$$

our result follows.

3.8.12 PETROVSKY'S THEOREM. If P is an elliptic operator of order m with analytic coefficients from ϑ_r to ϑ_s on $\Omega \subset R^n$, and if $u \in C^{\infty}(\Omega, r)$ is such that Pu is analytic, then u is analytic.

PROOF. Replacing P by $P^* \circ P$ if necessary, we may suppose that P is an operator from ϑ_r to itself. Let $f = Pu$ be analytic. It follows easily from theorem 3.8.10 that there is a constant $M > 0$ so that

$$|P^{\nu}f(x)| \leq M^{\nu+1}(\nu m)! \; x \in \Omega' \Subset \Omega.$$

and the proof then follows from theorem 3.8.9.

3.8.13 REMARKS. We indicate briefly how the proof of theorem 3.8.9 simplifies in the special case needed for theorem 3.8.12. We apply again (3.8.8) and (3.8.8'). Using now Cauchy's inequalities to a holomorphic extension of $f = Pu$ into the complex domain, we conclude that

$$|D^{\alpha}Pu| \leqq \text{const}\, \alpha!(mk\rho)^{-|\alpha|} \qquad \text{for} \quad \|x\| \leqq R - mk\rho,$$

which gives

$$\rho^{|\alpha'|} M_{(|\alpha|-1)\rho}(D^{\alpha'}Pu) \leqq \text{const}, \qquad \text{for} \quad \rho > 0 \text{ and all } \alpha.$$

This leads to the estimate

$$\rho^{|\alpha|} M_{|\alpha|\rho}(D^\alpha u) \leqq A^{|\alpha|+1},\ A = A(u),$$

and the proof is completed as before.

The main theorem 3.8.12 is a special case of results of PETROVSKY [1939] who considered also non-linear systems of differential equations. His proof is, however, very difficult. The main idea of the proof given here is contained in the paper of MORREY-NIRENBERG [1957]. The exposition is based on HÖRMANDER [1963].

§ 3.9 The finiteness theorem

Let V be an oriented C^∞ manifold and let ξ, η be C^∞ vector bundles on V. Let $\xi = (E, p, V), \eta = (F, q, V)$. Let P be a linear differential operator from ξ to η of order m.

We shall consider sections of $\xi, \eta, \ldots$ which are not necessarily continuous. A section of ξ is a map s: $V \to E$ such that $p \circ s =$ identity. We shall say that a section s is measurable if every point $a \in V$ has a neighbourhood U and an isomorphism τ: $\xi|U \to U \times C^r$ such that the section $\tau \circ s$ of the trivial bundle $U \times C^r$, considered as an r-tuple of functions, is measurable. We say that s is locally in H_m if U above can be chosen diffeomorphic to an open set Ω in R^n and $\tau \circ s$, considered as an r-tuple of functions on Ω, is in $H_m(\Omega)$. We shall use similar terminology for locally integrable sections and so on.

If P is a differential operator from ξ to η and s is a locally integrable section of ξ, we define Ps as the linear functional λ on $C_0^\infty(V, \eta')$ defined by *

$$\lambda(t') = \langle s, P't' \rangle_{\xi'},$$

where P' is the transpose of P. We shall say that Ps is locally integrable ($C^\infty, \ldots$) if there is a locally integrable ($C^\infty, \ldots$) section t of η so that

$$\lambda(t') = \langle s, P't' \rangle_{\xi'} = \langle t, t' \rangle_{\eta'} \qquad \text{for all} \quad t' \in C_0^\infty(V, \eta').$$

* Note that if s is a locally square integrable section of ξ, s' a locally square integrable section of ξ' and if one of them vanishes outside a compact subset of V, we can define

$$\langle s', s \rangle_\xi = \int_V B(s', s) \text{ as in § 3.3.}$$

The element t, if it exists, is uniquely determined (upto sets of measure zero), and we shall then identify Ps with t.

As an immediate consequence of theorems 3.7.7 and 3.8.12, we obtain the following two theorems.

3.9.1 THE REGULARITY THEOREM. Let ξ, η be C^∞ bundles on the oriented C^∞ manifold V and P an elliptic operator from ξ to η. If s is a locally square integrable section of ξ such that Ps is a C^∞ section of η, then s is equal (almost everywhere) to a C^∞ section of ξ.

3.9.2 THE ANALYTICITY THEOREM. Let V be a real analytic manifold, ξ, η analytic vector bundles on V. Let P be an elliptic operator from ξ to η with analytic coefficients. Then, if s is a locally square integrable section of ξ such that Ps is an analytic section of η, then s is equal (almost everywhere) to an analytic section of ξ.

If s is a section of the bundle ξ, we set again supp (s) = closure in V of the set $\{x \in V \mid s(x) \neq 0_x\}$, where 0_x stands for the zero element in the vector space E_x.

Let K be a compact subset of V. We set $H_m(K, \xi)$ = the set of sections s of ξ which are locally in H_m and for which supp $(s) \subset K$. Let $U_1', \ldots, U_h'$ be a finite number of coordinate neighbourhoods such that $\xi | U_j'$ is trivial, and let $K \subset \bigcup U_j'$. Let $U_j \Subset U_j'$ and let $K \subset \bigcup U_j$. Let $\tau_j \colon \xi | U_j' \to U_j' \times C^r$ be C^∞ isomorphisms, and let $\varphi_j \in C_0^\infty(U_j)$, $\sum \varphi_j(x) = 1$ for x in a neighbourhood of K. Then, if $s \in H_m(K, \xi)$, $\tau_j(\varphi_j s)$ can be looked upon as an r-tuple of functions on U_j. We suppose that U_j' is isomorphic to an open set Ω_j' in R^n. If $\psi_j \colon U_j' \to \Omega_j'$ is an isomorphism, let $\psi_j(U_j) = \Omega_j$. Then, the $\tau_j(\varphi_j s)$ can be looked upon as an r-tuple $s_j = \tau_j(\varphi_j s) \circ \psi_j^{-1}$ of functions on Ω_j and we have

$$s_j \in H_m(\Omega_j).$$

We set $\mathfrak{U} = \{U_1, \ldots, U_h\}$ and

$$|s|_{m,\mathfrak{U}}^2 = \sum_j |s_j|_m^2 .$$

With respect to this norm, $H_m(K, \xi)$ is a Hilbert space. Let

$$\mathscr{H} = \bigoplus H_m(\Omega_j).$$

The map

$$\eta: H_m(K, \xi) \to \mathscr{H},$$

given by

$$\eta(s) = \oplus s_j$$

is an isometry of $H_m(K, \xi)$ onto a closed subspace of $\mathscr{H}$.

Furthermore, we have

$$\tau_j(s|U_j) \circ \psi_j^{-1} = s_j' \in H_m(\Omega_j).$$

If we set

$$||s||_{m,\mathfrak{U}}^2 = \sum_j |s_j'|_m^2,$$

$H_m(K, \xi)$ is again a Hilbert space with respect to the norm $||s||_{m,\mathfrak{U}}$. One sees easily that $|s|_{m,\mathfrak{U}}$ and $||s||_{m,\mathfrak{U}}$ are equivalent norms on $H_m(K, \xi)$ and that different coverings $\mathfrak{U}$ and different partitions of unity $\{\varphi_j\}$ with the properties listed above give rise to equivalent norms $|\cdot|_{m,\mathfrak{U}}$.

Exactly as in proposition 3.4.3, one defines an *injection*

$$i_m: H_m(K, \xi) \to H_{m-1}(K, \xi), \qquad m \geqq 1.$$

Further lemma 3.5.4 gives us:

3.9.3 Rellich's lemma. The injection

$$i_m: H_m(K, \xi) \to H_{m-1}(K, \xi)$$

is completely continuous.

3.9.4 Proposition. For any continuous linear functional l on $H_0(K, \xi)$, there is a unique $s' \in H_0(K, \xi')$ such that

$$l(s) = \langle s', s \rangle_\xi \qquad \text{for all} \quad s \in H_0(K, \xi).$$

Proof. It is clear that s', if it exists, is unique. Hence it suffices to prove the following:

Let U be a coordinate neighbourhood such that $\xi|U$ is trivial, and let L be a compact subset of U. Then there is an $s' \in H_0(L, \xi')$ such that

$$l(s) = \langle s', s \rangle_\xi \qquad \text{for all} \quad s \in H_0(L, \xi).$$

Let

$$h: \xi|U \to U \times C^r$$

be an isomorphism, and let

$$h^*: \xi^*|U \to U \times C^r$$

be the corresponding isomorphism of the dual bundles. Let $\psi: U \to \Omega$ be an isomorphism of U onto an open set $\Omega \subset R^n$ and let

$$(s_1, \ldots, s_r) = h(s) \circ \psi^{-1}, \qquad s \in H_0(L, \xi).$$

Then the $s_j \in L^2(\Omega)$ and vanish outside $\psi(L)$. By a well known theorem of Riesz, there are $t_1, \ldots, t_r \in L^2(\Omega)$ vanishing outside $\psi(L)$, such that

$$l(s) = \int_\Omega \sum_{j=1}^{r} s_j t_j \mathrm{d}x_1 \wedge \ldots \wedge \mathrm{d}x_n, \qquad s \in H_0(L, \xi).$$

If we set

$$s' = (h^*)^{-1}(t_1 \circ \psi, \ldots, t_r \circ \psi) \otimes \psi^*(\mathrm{d}x_1 \wedge \ldots \wedge \mathrm{d}x_n),$$

we have $s' \in H_0(L, \xi')$ and

$$l(s) = \langle s', s \rangle_\xi \qquad \text{for all} \quad s \in H_0(L, \xi).$$

3.9.5 Remark. If P is a linear differential operator of order m from ξ to η, P defines a continuous linear map

$$P_K: H_m(K, \xi) \to H_0(K, \eta);$$

here K is a compact subset of V.

3.9.6 Proposition. If P is an elliptic differential operator from ξ to η, let

$$\mathscr{P} = \{s \in C^\infty(V, \xi) | Ps = 0\}.$$

Then a sequence $\{s_\nu\}$ of elements of $\mathscr{P}$ converges, together with all partial derivatives, uniformly on compact subsets of V (see remarks following 3.3.10) if and only if $\{s_\nu\}$ converges in $H_0(K, \xi)$ for any compact set $K \subset V$, i.e. if the sequence $\{s'_\nu\}$ defined by

$$s'_\nu(x) = \begin{cases} s_\nu(x) & \text{if} \quad x \in K, \\ 0 & \text{if} \quad x \notin K, \end{cases}$$

converges in $H_0(K, \xi)$.

PROOF. This follows from theorem 3.6.12.

3.9.7 PROPOSITION. Let $\mathscr{H}_1, \mathscr{H}_2$ be Hilbert spaces and A_1, A_2 two continuous linear maps $\mathscr{H}_1 \to \mathscr{H}_2$ with the following properties:
(a) A_1 is injective and $A_1(\mathscr{H}_1)$ is closed in $\mathscr{H}_2$;
(b) A_2 is completely continuous.

Then ker (A_1+A_2) is of finite dimension and $(A_1+A_2)(\mathscr{H}_1)$ is closed in $\mathscr{H}_2$.

PROOF. By the closed graph theorem,

$$A_1^{-1} = B: A_1(\mathscr{H}_1) \to \mathscr{H}_1$$

is continuous. Suppose that ker (A_1+A_2) is of infinite dimension. Then there exists an infinite orthonormal sequence $\{x_\nu\}$, $x_\nu \in \mathscr{H}_1$ such that

$$A_1(x_\nu) = -A_2(x_\nu).$$

Since $\|x_\nu\| = 1$, and A_2 is completely continuous, we can suppose, by passing to a subsequence, that $A_2(x_\nu)$ converges to an element y_0. Hence

$$A_1(x_\nu) \to -y_0,$$

so that

$$x_\nu = B(A_1(x_\nu)) \to -B(y_0) \quad \text{as} \quad \nu \to \infty,$$

contradicting our assumption that $\{x_\nu\}$ is orthonormal. Hence $\ker(A_1+A_2)$ is of finite dimension.

Let $N = \ker(A_1+A_2)$ and M be the orthogonal complement of N in $\mathscr{H}_1$. Let T be the restriction of A_1+A_2 to M. Then T is continuous and injective, and $T(M) = (A_1+A_2)(\mathscr{H}_1)$. To prove that this space is closed, it is sufficient to prove that $T^{-1}|T(M)$ is continuous. Let

$$y_\nu \in T(M), \qquad y_\nu \to 0.$$

Let

$$x_\nu \in M, \qquad T(x_\nu) = y_\nu.$$

If $\{x_\nu\}$ does not tend to zero, we may suppose that $\|x_\nu\| \geqq \rho > 0$. Then, if $x'_\nu = x_\nu/\|x_\nu\|$, we have

$$T(x'_\nu) = A_1(x'_\nu)+A_2(x'_\nu) \to 0,$$

so that, again by the complete continuity of A_2, we may suppose that

$A_2(x'_\nu)$, and hence also $A_1(x'_\nu)$ converges. Hence, since B is continuous, $x'_\nu = BA_1(x'_\nu)$ converges, say to x'_0. Then clearly $\|x'_0\| = 1$. On the other hand

$$T(x'_0) = \lim T(x'_\nu) = 0,$$

so that $x'_0 \in N$. Since $x'_\nu \in M$, we have $x'_0 \in M$. This implies that $x'_0 = 0$, a contradiction.

3.9.8 THE FINITENESS THEOREM. Let V be an oriented C^∞ manifold and ξ, η be C^∞ vector bundles on V. Let P be an elliptic operator from ξ to η of order $m \geqq 1$ and K a compact subset of V. Then the map

$$P_K: H_m(K, \xi) \to H_0(K, \eta)$$

has a finite dimensional kernel and a closed image.

PROOF. Let

$$i_m: H_m(K, \xi) \to H_{m-1}(K, \xi)$$

be the natural injection. Let

$$\mathscr{H}_1 = H_m(K, \xi), \qquad \mathscr{H}_2 = H_0(K, \eta) \oplus H_{m-1}(K, \xi).$$

Let $A_1; A_2: \mathscr{H}_1 \to \mathscr{H}_2$ be defined as follows:

$$A_1(s) = P_K s \oplus i_m(s); A_2(s) = 0 \oplus -i_m(s).$$

Clearly A_1 is an injection and A_2 is completely continuous (lemma 3.9.3). Moreover, if U is a coordinate neighbourhood on V and $\varphi \in C_0^\infty(U)$, we obtain, from (3.6.11)

$$|\varphi s|_{m, \mathfrak{U}} \leqq \text{const}\, \{|P(\varphi s)|_{0, \mathfrak{U}} + |\varphi s|_{0, \mathfrak{U}}\} \leqq \text{const}\, \{|\varphi P s|_{0, \mathfrak{U}} + |s|_{m-1, \mathfrak{U}}\}$$

with respect to any fixed covering $\mathfrak{U}$ of K. This gives

$$|s|_{m, \mathfrak{U}} \leqq \text{const}\, \{|Ps|_{0, \mathfrak{U}} + |s|_{m-1, \mathfrak{U}}\}.$$

It follows that

$$\text{const}\, \|A_1(s)\|_{\mathscr{H}_2} \geqq \|s\|_{\mathscr{H}_1}, \qquad s \in \mathscr{H}_1,$$

so that $A_1(\mathscr{H}_1)$ is closed. By proposition 3.9.7,

$$\ker (A_1 + A_2) = \ker P_K$$

is finite dimensional and $(A_1+A_2)(\mathscr{H}_1) = P_K(H_m(K, \xi)) \oplus \{0\}$ is closed. The result follows.

3.9.9 PROPOSITION. Let V be an oriented C^∞ manifold, ξ, η vector bundles and P an elliptic operator of order m from ξ to η. Let K be a compact subset of V. Then, if $t_0 \in H_0(K, \eta)$ is such that $\langle t', t_0\rangle_\eta = 0$ for all $t' \in H_0(K, \eta')$ with $P't' = 0$ on $\mathring{K}$, there exists an $s_0 \in H_m(K, \xi)$ with $Ps_0 = t_0$.

PROOF. Let

$$N = \{t \in H_0(K, \eta) \mid \langle t', t\rangle_\eta = 0 \quad \text{for all} \quad t' \in H_0(K, \eta') \quad \text{with} \quad P't' = 0 \quad \text{on} \quad \mathring{K}\}.$$

The equation $P't' = 0$ on $\mathring{K}$ means that $\langle Ps, t'\rangle_{\eta'} = 0$ for all $s \in C_0^\infty(\mathring{K}, \xi)$. Let l be a continuous linear functional on $H_0(K, \eta)$ which vanishes on $P_K(H_m(K, \xi))$. We have to show, since $P_K(H_m(K, \xi))$ is closed, that $l(t_0) = 0$. Now, there is $t_0' \in H_0(K, \eta')$ with

$$l(t) = \langle t_0', t\rangle_\eta$$

for all $t \in H_0(K, \eta)$ (proposition 3.9.4). Since l vanishes on $P_K(H_m(K, \xi))$,

$$l(Ps) = \langle t_0', Ps\rangle_\eta = 0$$

for all $s \in C_0^\infty(\mathring{K}, \xi)$, so that $P't_0' = 0$ on $\mathring{K}$. By definition of N,

$$\langle t_0', t\rangle_\eta = l(t) = 0 \qquad \text{for} \quad t \in N,$$

which proves the proposition.

3.9.10 PROPOSITION. If, in addition, V is compact, and $K = V$, we have

$$P_V(H_m(V, \xi)) = \{t \in H_0(K, \eta) \mid \langle t', t\rangle_\eta = 0 \quad \text{for all} \quad t' \in H_0(V, \eta') \quad \text{with} \quad P't' = 0\}.$$

PROOF. If we denote the space on the right above by N, theorem 3.9.9 implies that

$$P_V(H_m(V, \xi)) \supset N.$$

On the other hand, $\langle Ps, t'\rangle_{\eta'} = 0$ whenever $P't' = 0$, so that $N \supset P_V(H_m(V, \xi))$.

3.9.11 PROPOSITION. Let V be a compact oriented C^∞ manifold and ξ, η be C^∞ vector bundles with rank (ξ) = rank (η). Let P be an elliptic operator from ξ to η. Then $C^\infty(V, \eta)/P(C^\infty(V, \xi))$ is of finite dimension.

PROOF. Consider the operator

$$P_V\colon H_m(V, \xi) \to H_0(V, \eta);$$

let M be its image. By proposition 3.9.10,

$$M = \{t \in H_0(V, \eta) | \langle t', t\rangle_\eta = 0 \quad \text{for all} \quad t' \in H_0(V, \eta') \quad \text{with} \quad P't' = 0\}.$$

Hence, if

$$P'_V\colon H_m(V, \eta') \to H_0(V, \xi')$$

is the operator corresponding to the formal transpose P' from η' to ξ', then

$$\text{cokernel}\,(P_V) \simeq \text{kernel}\,(P'_V).$$

Since rank (ξ) = rank (η), P' is elliptic (remark 3.3.22), so that, by proposition 3.9.8, cokernel (P_V) is finite dimensional.
Also

$$M \cap C^\infty(V, \eta) = P(H_m(V, \xi)) \cap C^\infty(V, \eta) = P(C^\infty(V, \xi))$$

by theorem 3.9.1. Since M has finite codimension in $H_0(V, \eta)$, it follows that $P(C^\infty(V, \xi))$ has finite codimension in $C^\infty(V, \eta)$.

The results of this section extend to non-orientable manifolds. One has only to replace $\mathscr{E}^n$ in the definition of the bundle ξ' in § 3.3 by the socalled "volume bundle". This is the bundle obtained from $\mathscr{E}^n$ by replacing its transition functions g_{ij} by $|g_{ij}|$.

Furthermore, Proposition 3.9.6 can be proved simply by combining the regularity theorem with the closed graph theorem for Fréchet spaces. This has the advantage of being applicable to all linear differential operators for which the regularity theorem is true, even if the analogue of the Friedrichs inequality fails.

§ 3.10 The approximation theorem and its application to open Riemann surfaces

We shall assume, throughout this section, that V is connected.

3.10.1 NOTATION. Let V be a C^∞ manifold with a countable base and let S be a subset of V. We denote by $\mathscr{J}(S)$ the union of S with the relatively compact connected components of $V-S$.

We shall need several properties of the sets $\mathscr{J}(S)$.

3.10.2 PROPOSITION. If $S_1 \subset S_2$, then

$$\mathscr{J}(S_1) \subset \mathscr{J}(S_2);$$

further

$$\mathscr{J}(\mathscr{J}(S)) = \mathscr{J}(S).$$

PROOF. If C is a relatively compact connected component of $V-S_1$, then $C-S_2$ is a union of connected components of $V-S_2$. To see this, we have only to remark that if C' is a connected component of $V-S_2$ not contained in C and $C \cap C' \neq \emptyset$, then $C \cup C'$ is a connected subset of $V-S_1$ containing C properly, which is not possible since C is a component. Hence $C-S_2$ is a union of components of $V-S_2$; since C is relatively compact, this implies that $C \subset \mathscr{J}(S_2)$.

3.10.3 PROPOSITION. If S is closed, $\mathscr{J}(S)$ is closed. If K is compact, $\mathscr{J}(K)$ is also compact.

PROOF. If S is closed, any component of $V-S$ is open. Hence $V-\mathscr{J}(S)$, being the union of those components of $V-S$ which are not relatively compact,is again open.

Let U be a relatively compact open set contaning K. Let $U_1, \ldots U_h$ be connected open sets covering ∂U, $U_j \cap K = \emptyset$. Clearly, any connected component of $V-K$ not contained in U must contain at least one of the U_j. Hence, there are only finitely many relatively compact components of $V-K$ not contained in U, and it follows that $\mathscr{J}(K)$ is relatively compact.

For the next result, we shall need the following proposition.

3.10.4 PROPOSITION. Let X be a locally compact Hausdorff topological space, and let K_0 be a compact connected component of X. Then K_0 has a fundamental system of neighbourhoods which are simultaneously open and closed in X.

PROOF. Replacing X by a compact neighbourhood of K_0 if necessary, we may suppose that X is compact.

Let $\mathscr{F}$ be the family of neighbourhoods N of K_0 which are simultaneously open and closed. ($\mathscr{F} \neq \emptyset$ since $X \in \mathscr{F}$). Let

$$K = \bigcap_{N \in \mathscr{F}} N.$$

Clearly K is closed, hence compact. Further $\mathscr{F}$ is closed under finite intersections. Hence K has a fundamental system of neighbourhoods which are elements of $\mathscr{F}$. Hence we have only to show that $K = K_0$. Since $K_0 \subset K$ and K_0 is a connected component, it suffices to show that K is connected.

Suppose K were not connected. Then

$$K = A_0 \cup A_1,$$

where A_0, A_1 are closed subsets of K (hence compact), $A_0 \cap A_1 = \emptyset$, and neither A_0 nor A_1 is empty. Since A_0, A_1 are disjoint, compact sets, we can find open sets U_0, U_1 with $A_j \subset U_j$, $U_0 \cap U_1 = \emptyset$. Let $U = U_0 \cup U_1$. Since

$$\bigcap_{N \in \mathscr{F}} (N \cap (X - U)) = \emptyset,$$

and $\mathscr{F}$ is closed under finite intersections, and since $X - U$ is compact, it follows that there is an

$$N \in \mathscr{F}, \qquad N \cap (X - U) = \emptyset,$$

i.e. $N \subset U$. Clearly K_0 is contained either in U_0 or U_1, say $K_0 \subset U_0$. But then

$$N \cap U_0 = N \cap (X - U_1)$$

is both open and closed, so that

$$K \subset N \cap U_0 \subset U_0,$$

a contradiction since

$$K \cap U_1 \supset A_1 \neq \emptyset.$$

This proves the proposition.

3.10.5 PROPOSITION. If S is an open subset of V, then $\mathscr{I}(S)$ is also open.

PROOF. Let K_0 be a relatively compact component of $V-S$. Then K_0 is compact. Let N be a neighbourhood of K_0 which is compact and open in $V-S$. Then $V-S-N$ is closed in $V-S$, hence in V, so that $S \cup N$ is open in V. Clearly $N \subset \mathscr{J}(S)$, and it follows that $\mathscr{J}(S)$ is open.

3.10.6 PROPOSITION. Let K be a compact set and let $K = \mathscr{J}(K)$. Then K has a fundamental system of open (or compact) neighbourhoods S with $S = \mathscr{J}(S)$.

PROOF. We may suppose that V is connected. It is easily shown that K has a fundamental system of open neighbourhoods U such that $V-U$ has only finitely many connected components. Let $U' = \mathscr{J}(U)$, and let C_0 be a compact component of $V-U$. Then $C_0 \subset V-K$, and so $C_0 \subset U_0$, where U_0 is an (open) connected component of $V-K$. Since $K = \mathscr{J}(K)$, U_0 is not relatively compact, so that $\partial U' \cap U_0 \neq \emptyset$. Let γ_0 be a simple arc joining a point of C_0 to a point of $\partial U'$ and lying in U_0. We construct such a curve γ_j for each compact component C_j of $V-U$. Then clearly, if

$$S = U - \bigcup_j \gamma_j,$$

S is open, and we have $S = \mathscr{J}(S)$.

If S is an open neighbourhood of K with $S = \mathscr{J}(S)$, and L is a compact neighbourhood of $K, L \subset S$, then, by proposition 3.10.2, $L' = \mathscr{J}(L) \subset S$ and by proposition 3.10.3, L' is a compact neighbourhood of K. This proves the proposition.

3.10.7 THE APPROXIMATION THEOREM OF MALGRANGE-LAX. Let V be an oriented real analytic manifold and ξ, η be real analytic vector bundles on V with rank (ξ) = rank (η). Let P be an elliptic operator of order m from ξ to η with analytic coefficients. Let U be an open subset of V such that $V-U$ has no compact connected components. Then, any $u \in C^\infty(U, \xi)$ with $Pu = 0$ on U is the limit together with all partial derivatives, uniformly on compact subsets of U, of sections $u_\nu \in C^\infty(V, \xi)$ with $Pu_\nu = 0$ on V.

PROOF. Let K be a compact set in U. We can suppose, by replacing K by $\mathscr{J}(K)$, that $K = \mathscr{J}(K)$, note that by propositions 3.10.2, 3 $\mathscr{J}(K)$ is compact and $\subset U$.

Let L be a compact set in V such that $K \subset \mathring{L}$, and let

$$\mathscr{P}(L) = \{s \in H_0(L, \xi) | Ps = 0 \quad \text{on} \quad \mathring{L}\},$$

and $\mathscr{S}(K)$ be the restrictions to $H_0(K, \xi)$ of sections $s \in C^\infty(N, \xi)$ with $Ps = 0$ on N, N being a neighbourhood of K which may depend on s. Let

$$\rho: \mathscr{P}(L) \to H_0(K, \xi)$$

be the map

$$\rho(s)(x) = \begin{cases} s(x) & \text{if} \quad x \in K, \\ 0 & \text{if} \quad x \notin K, \end{cases}$$

and let $M = \rho(\mathscr{P}(L))$. Clearly $M \subset \mathscr{S}(K)$. We first prove:

3.10.8 THEOREM. *M is dense in $\mathscr{S}(K)$.*

PROOF. Let l be a continuous linear functional on $H_0(K, \xi)$ such that $l|M = 0$. We have to prove that $l|\mathscr{S}(K) = 0$. Now, by proposition 3.9.4, there is an $s_0' \in H_0(K, \xi')$ such that

$$l(s) = \langle s_0', s \rangle_\xi, \qquad s \in H_0(K, \xi).$$

Define $s' \in H_0(L, \xi')$ by

$$s'(x) = \begin{cases} s_0'(x), & x \in K, \\ 0, & x \notin K. \end{cases}$$

We claim that

$$\langle s', u \rangle_\xi = 0$$

if $u \in H_0(L, \xi)$ and

$$Pu = 0 \quad \text{on} \quad \mathring{L}.$$

In fact

$$\langle s', u \rangle_\xi = \langle s_0', \rho(u) \rangle_\xi = l(\rho(u)) = 0$$

since $l|M = 0$.

Hence, by proposition 3.9.9, there exists $t' \in H_m(L, \eta')$ such that $P't' = s'$. Now, $s'(x) = 0$ if $x \notin K$. Hence $P't' = 0$ on $V-K$. Since P' is an elliptic operator with analytic coefficients (remark 3.3.22), it follows from theorem 3.9.2 that t' is analytic on $V-K$. Now $t'(x) = 0$ if $x \notin L$. Furthermore, no connected component of $V-K$ is contained in L since $K = \mathscr{J}(K)$. Hence t' vanishes on a non-empty open subset of any connected component of $V-K$, and being analytic, $t'(x) = 0$ for any $x \in V-K$.

If $s \in \mathscr{S}(K)$ and N is a neighbourhood of K in which s is defined and $Ps = 0$, we have

$$l(s) = \langle s_0', s\rangle_\xi = \langle P't', s\rangle_\xi = \int_N B(P't', s)$$

$$\text{(since } P't' = 0 \text{ on } N-K)$$

$$= \int_N B(t', Ps)$$

(since supp (t') is compact in N) and this equals 0 since $Ps = 0$. Thus $l|\mathscr{S}(K) = 0$, and the theorem is proved.

Thus, by theorem 3.9.1 there exists a sequence $u_\nu \in C^\infty(\mathring{L}, \xi)$, $Pu_\nu = 0$, such that $u_\nu \to u$ in $H_0(K, \xi)$. (By proposition 3.9.6, u_ν converges to u uniformly on $\mathring{K}$ together with all its derivatives.)

Let $\{K_\nu\}$ be a sequence of compact subsets of V with

$$K \subset \mathring{K}_1, \qquad K_\nu \subset \mathring{K}_{\nu+1}, \qquad K_\nu = \mathscr{J}(K_\nu).$$

(It is easily seen that such a sequence exists.) If $\varepsilon > 0$ is given and $s \in \mathscr{S}(K)$, we obtain from theorem 3.10.8 the existence of

$$s_\nu \in C^\infty(\mathring{K}_{\nu+1}, \xi), \qquad Ps_\nu = 0$$

such that

$$|s_1 - s|^K < \varepsilon/2, \qquad |s_{\nu+1} - s_\nu|^{K_\nu} < \varepsilon/2^\nu;$$

here $|\ldots|^{K_\nu}$ denotes a norm defining the topology of $H_0(K_\nu, \xi)$ with $|s|^{K_\nu} \leqq |s|^{K_{\nu+1}}$ for $s \in H_0(K_{\nu+1}, \xi)$. Then the series

$$u = s_\nu + \sum_{\mu=\nu+1}^{\infty} (s_\mu - s_{\mu-1})$$

converges in $H_0(K_\nu, \xi)$, and the sum is independent of ν. Moreover, by proposition 3.9.6, $u \in C^\infty(V, \xi)$ and $Pu = 0$. Clearly $|u-s|^K < \varepsilon$. Thus there is a sequence $\{u_N\}$ in $C^\infty(V, \xi)$, $Pu_N = 0$ converging to s in $H_0(K, \xi)$. The theorem follows at once from proposition 3.9.6.

From the above theorem and proposition 3.10.6 we obtain immediately:

3.10.9 COROLLARY. Let K be a compact subset of V with $K = \mathscr{J}(K)$. Then, with the notation of theorem 3.10.8, if u is a solution of the equation $Pu = 0$ in a neighbourhood of K, u can be approximated, together with all its derivatives, by solutions of the equation $Ps = 0$ on V, uniformly on K.

3.10.10 REMARK. It can be proved that the condition that $U = \mathscr{J}(U)$ in theorem 3.10.8 is also necessary for the approximation theorem to hold. The proof uses, however, the existence theory for the equation $Pu = f$ which we have not treated. See MALGRANGE [1955/56].

Let now V be a complex manifold of dimension n and $\mathscr{E}^{p,q}$ the bundle of forms of type (p, q) on V. We have already remarked (example 3.3.15, (b)) that the differential operator $\bar{\partial}$ from $\mathscr{E}^{p,0}$ to $\mathscr{E}^{p,1}$ is elliptic; in particular, $\bar{\partial}$ from ϑ_1 to $\mathscr{E}^{0,1}$ is elliptic. Now $\operatorname{rank}(\vartheta_1) = 1$ and $\operatorname{rank}(\mathscr{E}^{0,1}) = n$. If $n = 1$, we can therefore apply theorem 3.10.7, and we obtain the following theorem:

3.10.11 THE RUNGE THEOREM FOR OPEN RIEMANN SURFACES: BEHNKE-STEIN. Let V be a connected complex manifold of complex dimension 1, with a countable base, i.e. an open Riemann surface Let U be an open subset of V such that $V - U$ has no compact connected components. Then, any holomorphic function on U is the limit of holomorphic functions on V, uniformly on compact subsets of U.

One of the applications of this theorem which has far reaching consequences is theorem 3.10.13 below. It proves, in particular, the conjecture of Carathéodory that on any open Riemann surface, there exist non-constant holomorphic functions.

3.10.12 DEFINITION. Let V be a complex manifold of dimension n and $\mathscr{H} = \mathscr{H}(V)$ the ring of holomorphic functions on V. V is called a Stein manifold if the following three conditions are satisfied.

(a) $\mathscr{H}$ separates the points of V, i.e. if $a, b \in V$, $a \neq b$, there is $f \in \mathscr{H}$ with $f(a) \neq f(b)$.

(b) If $a \in V$, there are $f_1, \ldots, f_n \in \mathscr{H}$ such that the map $f\colon V \to C^n$ defined by $f_1, \ldots, f_n$ is an isomorphism of a neighbourhood of a onto an open set in C^n, i.e. $f_1, \ldots, f_n$ give local coordinates in a neighbourhood of a.

(c) For any compact set $K \subset V$, the set

$$\hat{K} = \hat{K}_{\mathscr{H}} = \{x \in V \mid |f(x)| \leqq \sup_{y \in K} |f(y)| \quad \text{for all} \quad f \in \mathscr{H}\}$$

is compact.

3.10.13 THEOREM (BEHNKE-STEIN). Every open Riemann surface is a Stein manifold.

PROOF. Let $a, a' \in V$, V being the given open Riemann surface. Let (U, φ), (U', φ') be coordinate systems with $U \cap U' = \emptyset$,

$$\varphi(U) = \{z \in C \mid |z| < 1\} = \varphi'(U') \text{ and } \varphi(a) = \varphi'(a') = 0.$$

If

$$D = \{x \in U \mid |\varphi(x)| < r < 1\},$$

$$D' = \{x \in U' \mid |\varphi'(x')| < r < 1\},$$

then one sees at once that $V-D$, $V-D'$ and $V-D-D'$ are connected. Hence, if $\Omega = D \cup D'$, $V-\Omega$ has no compact connected component. By theorem 3.10.11, the function u on Ω defined by

$$u(x) = \begin{cases} 0, & x \in D, \\ 1, & x \in D', \end{cases}$$

can be approximated, uniformly on $K' = \{a\} \cup \{a'\}$ by elements $f \in \mathscr{H}$. In particular, there is $f \in \mathscr{H}$ with $|f(a)| < \frac{1}{2}$, $|f(a')| > \frac{1}{2}$. Hence $\mathscr{H}$ separates points.

If $a \in V$, and (U, φ) is a coordinate neighbourhood as above and

$$D = \{x \in U \mid |\varphi(x)| < r < 1\}, \qquad K = \{x \in U \mid |\varphi(x)| \leqq r' < r\},$$

then $V-D$ is connected. Again by theorem 3.10.8, there is $f \in \mathscr{H}$ with

$$\sup_{y \in K} |f(y) - \varphi(y)| < \varepsilon.$$

If ε is small enough, $(\mathrm{d}f)(a) \neq 0$, and it follows that f is a local homeomorphism in the neighbourhood of a, which shows that f gives local coordinates at a. As for (c) in definition 3.10.12, we shall show that

$$\hat{K}_{\mathscr{H}} = \mathscr{I}(K).$$

First, if U is a relatively compact connected component of $V-K$, we have $\partial U \subset K$. Hence, it follows from the maximum principle that $U \subset \hat{K}_{\mathscr{H}}$, so that

$$\mathscr{J}(K) \subset \hat{K}_{\mathscr{H}}.$$

Let now $a \notin \mathscr{J}(K)$, and let $L = \{a\} \cup \mathscr{J}(K)$. We see immediately that $L = \mathscr{J}(L)$. Hence, by corollary 3.10.9 applied to $\bar{\partial}$ from ϑ_1 to $\mathscr{E}^{0,1}$, there is $f \in \mathscr{H}$ with

$$\sup_{y \in L} |f(y)-u(y)| < \tfrac{1}{2},$$

where

$$u(y) = \begin{cases} 1 & \text{if } y \text{ is in a neighbourhood of } a, \\ 0 & \text{if } y \text{ is in a neighbourhood of } \mathscr{J}(K). \end{cases}$$

Then

$$|f(a)| > \sup_{y \in K} |f(y)|,$$

so that $a \notin \hat{K}_{\mathscr{H}}$. Hence

$$\hat{K}_{\mathscr{H}} \subset \mathscr{J}(K),$$

and the theorem is proved.

The main theorem 3.10.8 is due to MALGRANGE [1955/56] and LAX [1956]. The application to open Riemann surfaces is as in MALGRANGE [1955/56]. The original treatment of BEHNKE-STEIN [1948] is completely different. The method of Behnke-Stein is rather difficult and uses the theory of compact Riemann surfaces in great detail, but has the advantage of giving, at the same time, solutions to the so called ‘first and second problems of Cousin’ (theorems of Mittag-Leffler and Weierstrass). Actually, the first Cousin problem can be solved rather simply using the generalization of the Behnke-Stein theorem to sections of a holomorphic vector bundle, a result which follows immediately from the Malgrange-Lax theorem.

References

Asterisks refer to books

ABRAHAM, R., 1963, *Transversality in manifolds of mappings*, Bull. Am. Math. Soc. **69**, 470–474.

ATIYAH, M. F., 1962, *Immersions and imbeddings of manifolds*, Topology **1**, 125–132.

BEHNKE, H. AND K. STEIN, 1948, *Entwicklung analytischer Funktionen auf Riemannschen Flächen*, Math. Annalen **120**, 430–461.

* BOURBAKI, N., 1952, ALGÈBRE, Ch. VII, *Modules sur les anneaux principaux*, Hermann, Paris.

* BOURBAKI, N., 1965, *Topologie générale*, Ch. I, *Structures topologiques, 4e édition*, Hermann, Paris.

* BOURBAKI, N., 1958, *Topologie générale*, Ch. IX, *Utilisation des nombres réels en topologie générale*, Hermann, Paris.

BISHOP, E., 1961, *Mappings of partially analytic spaces*, Am. J. Math. **83**, 209–242.

* CARTAN, E., 1958, *Leçons sur les invariants intégraux*, Hermann, Paris.

CARTAN, H., 1957, *Variétés analytiques réelles et variétés analytiques complexes*, Bull. Soc. Math. de France **85**, 77–99.

CARTAN, H., 1961/62, *Séminaire E.N.S.: Topologie différentielle.*

CHEVALLEY, C., 1944, *Theory of Lie groups*, Princeton Univ. Press.

DIEUDONNÉ, J., 1944, *Une généralisation des espaces compacts*, J. Math. Pure Appl. **23**, 65–76.

FRIEDRICHS K. O., 1953, *On the differentiability of the solutions of linear elliptic differential equations*, Commun. Pure Appl. Math. **6**, 299–325.

FUCHS, W. H. J., 1964, *On the eigenvalues of an integral equation arising in the theory of band limited signals*, J. Math. Analysis Applications **9**, 317–330.

GÅRDING, L., 1953, *Dirichlet's problem for linear elliptic partial differential equations*, Math. Scandinavica **1**, 55–72.

GLAESER, G., 1958, *Etudes de quelques algèbres Tayloriennes*, Jour d'Analyse (Jerusalem) **6**, 1–124.

GRAUERT, H., 1958, *On Levi's problem and the imbedding of real analytic manifolds*, Annals Math. **68**, 460–472.

HAEFLIGER, A., 1961, *Plongements différentiables de variétés dans variétés*, Commentarii Math. Helvetici **36**, 47–82.

* HERVE, M., 1963, *Several Complex Variables*, Tata Institute of Fundamental Research, Bombay and Oxford Univ. Press.

HIRSCH, M. W., 1961, *On imbedding differentiable manifolds in Euclidean space*, Annals Math. **73**, 566–571.

HOPF, H., 1948, *Zur Topologie der komplexen Mannigfaltigkeiten*, Studies and Essays presented to R. Courant (Interscience N.Y.), 167–185.

* HÖRMANDER, L., 1963, *Linear partial differential operators*, Springer, Berlin.

HÖRMANDER, L., 1964, *The Frobenius Nirenberg theorem*, Arkiv för Mathematik **5**, 425–432.

HÖRMANDER, L., 1965, *Pseudo differential operators*, Commun. Pure Appl. Math. **18**, 501–517.

* HÖRMANDER, L., 1966, *An introduction to complex analysis in several variables*, Van Nostrand, Princeton.

* HUREWICZ, W., AND H. WALLMAN, 1948, *Dimension Theory*, Princeton Univ. Press.

JOHN, F., 1955, *Plane waves and spherical means applied to partial differential equations*, Interscience N.Y.

KERVAIRE, M., 1960, *A manifold which does not admit any differentiable structure*, Commentarii Math. Helvetici **34**, 257–270.

KODAIRA, K. AND D. C. SPENCER, 1958, *On deformations of complex analytic structures, Part I & II*, Annals Math. **67**, 328–466.

KOHN, J. J., 1963, *Harmonic integrals on strongly pseudo-convex manifolds, I*, Annals Math. **78**, 206–213.

* KOSZUL, J. L., 1960, *Lectures on fibre bundles and differential geometry*, Tata Institute of Fundamental Research, Bombay.

KOTAKÉ, T. AND M. S. NARASIMHAN, 1962, *Regularity theorems for fractional powers of a linear elliptic operator*, Bull. Soc. Math. France **90**, 449–471.

LAX, P., 1955, *On Cauchy's problem for hyperbolic equations and the differentiability of solutions of elliptic equations*, Commun. Pure Appl. Math. **8**, 615–633.

LAX, P., 1956, *A stability theorem for abstract differential equations and its application to the study of the local behaviour of solutions of elliptic equations*, Commun. Pure Appl. Math. **9**, 747–766.

MALGRANGE, B., 1955/56, *Existence et approximation des solutions des équations aux dérivées partielles et des équations de convolution*, Annales de l'Institut Fourier **6**, 271–355.

* MALGRANGE, B., 1966, *Ideals of differentiable functions*, Tata Institute of Fundamental Research, Bombay and Oxford Univ. Press.

MALGRANGE, B., 1969, *Sur l'intégrabilité des structures presque-complexes*, Symposia Mathematica, Istituto di Alta Matematica, Academic Press, London and New York. pp. 289–296.

MILNOR, J., 1956, *On manifolds which are homeomorphic to the 7-sphere*, Annals Math. **64**, 399–405.

MORREY, C. B. AND L. NIRENBERG, 1957, *On the analyticity of the solutions of linear elliptic systems of partial differential equations*, Commun. Pure Appl. Math. **10**, 271–290.

MORSE, A. P., 1939, *The behaviour of a function on its critical set*, Annals Math. **40**, 62–70.

NARASIMHAN, R., 1960, *Imbedding of holomorphically complete complex spaces* Am. J. Math. **82**, 917–934.

NEWLANDER, A., AND L. NIRENBERG, *Complex analytic co-ordinates in almost complex manifolds*, Annals Math. **65**, 391–404.

NIRENBERG, L., 1955, *Remarks on strongly elliptic partial differential equations,* Commun. Pure Appl. Math. **8**, 648–674.

NIRENBERG, L., 1970, *Pseudo-differential operators,* Amer. Math. Soc. Symposium on Pure Math. Vol. XVI, Providence R.I. pp. 149–167.

* NOMIZU, K., 1956, *Lie groups and differential geometry,* Publications of the math. Society of Japan.

OKA, K., 1936, *Sur les fonctions analytiques de plusieurs variables, I. Domaines convexes par rapport aux fonctions rationelles,* J. Science, Hiroshima Univ. **6**, 245–255.

PAPY, G., 1956, *Sur la définition intrinsèque des vecteurs tangents à une variété de classe C^r lorsque $1 \leqq r < \infty$,* C. R. Acad. Sci. Paris **242**, 1573–1575.

PEETRE, J., 1960, *Rectifications à l'article „Une caractérisation abstraite des opérateurs différentiels",* Math. Scandinavica **8**, 116–120.

PETROVSKY, I. G., 1939, *Sur l'analyticité des systemes d'equations différentielles,* Math. Sbornik **5**, 3–68.

RELLICH, F., 1930, *Ein Satz über mittlere Konvergenz,* Göttinger Nachrichten, 30–35.

SARD, A., 1942, *The measure of critical values of differentiable maps,* Bull. Am. Math. Soc. **45**, 883–890.

SCHWARTZ, J., 1954, *The formula for change in variables in a multiple integral,* Am. Math. Monthly **61**, 81–85.

* SCHWARTZ, L., 1950/51, *Théorie des distributions,* Vol. 1 & 2, Hermann, Paris.

SCHWARTZ, L., 1963/64, *Les travaux de Seeley sur les opérateurs intégraux singuliers sur une variété,* Séminaire Bourbaki, 1963/64, Exposé 269.

SERRE, J. P., 1953/54, *Exposé* 18 *in Séminaire H. Cartan.*

SOBOLEV, S. L., 1938, *Sur un théorème d'analyse fonctionelle,* Math. Sbornik **4**, 471–496.

THOM, R., 1956, *Un lemme sur les applications différentiables,* Bol. Soc. Mat. Mexicana, 59–71.

WALL, C. T. C., 1965, *All 3-manifolds imbed in 5-space,* Bull. Am. Math. Soc. **71**, 564–567.

WEIL, A., 1952, *Sur les théorèmes de deRham,* Commentarii Math. Helvetici **26**, 119–145.

WEYL, H., 1916, *Über die Gleichverteilung von Zahlen mod Eins,* Math. Annalen **77**, 313–352.

WHITNEY, H., 1934, *Analytic extensions of differentiable functions defined in closed sets,* Trans. Am. Math. Soc. **36**, 63–89.

WHITNEY, H., 1935, *A function not constant on a connected set of critical points,* Duke Math. J. **1** 514–517.

WHITNEY, H., 1936, *Differentiable manifolds,* Annals Math. **37**, 645–680.

WHITNEY, H., 1944a, *The self-intersections of a smooth n-manifold in 2n-space,* Annals Math. **45**, 220–246.

WHITNEY, H., 1944b, *The singularities of a smooth n-manifold in $(2n-1)$-space,* Annals Math. **45**, 247–293.

* WHITNEY, H., 1957, *Geometric integration theory,* Princeton Univ. Press.

Subject Index

Printed and bound by CPI Group (UK) Ltd, Croydon, CR0 4YY
03/10/2024
01040430-0009